Augmented Reality with Unity AR Foundation

A practical guide to cross-platform AR development with Unity 2020 and later versions

Jonathan Linowes

BIRMINGHAM—MUMBAI

Augmented Reality with Unity AR Foundation

Copyright © 2021 Packt Publishing

Group Product Manager: Pavan Ramchandani
Senior Editor: Hayden Edwards
Content Development Editor: Aamir Ahmed
Technical Editor: Saurabh Kadave
Copy Editor: Safis Editing
Project Coordinator: Manthan Patel
Proofreader: Safis Editing
Indexer: Tejal Daruwale Soni
Production Designer: Sinhayna Bais

First published: August 2021

Production reference: 1130821

Published by Packt Publishing Ltd.
Livery Place
35 Livery Street
Birmingham
B3 2PB, UK.

ISBN 978-1-83898-259-1

www.packt.com

*Yes, I love my wife, Lisa, and our four amazing children, but this book
I dedicate to my dog, Coder, a large, sweet, and boisterous Doberman
Shepherd who keeps trying to pull me from my desk to play frisbee outside
in the real world.*

– Jon

Contributors

About the author

Jonathan Linowes is a VR/AR enthusiast, Unity, and full stack developer, entrepreneur, certified Unity instructor, and the owner of Parkerhill XR Studio, an immersive media, applications, and game developer. Jonathan has a bachelor of fine arts degree from Syracuse University, a master of science degree from the MIT Media Lab, and has held technical leadership positions at Autodesk and other companies. He has authored multiple books on VR and AR published by Packt Publishing.

About the reviewers

David Cantón Nadales is a software engineer from Seville, Spain, with over 15 years of experience. He specializes in Firebase and the development of mobile apps and video games, and is experienced in VR/AR with Oculus, Hololens, HTC Vive, DayDream, and LeapMotion. He was the ambassador of the Samsung development community "Samsung Dev Spain," and the organizer of the "Google Developers Group Seville" community He has worked on, and brought to fruition, more than 50 projects during his professional career. As a social entrepreneur, Grita stands out, a social network that emerged during the period of COVID-19 confinement that allowed people to talk to others in the same circumstances and help each other psychologically.

Rohit Gonsalves is connecting computer graphics, video, media, and devices to create beautiful **Extended Reality** (**XR**) workflows for PC and mobiles. He has 17+ years of experience working with computer graphics, specifically, C++, Windows, and DirectX. He is also the author of an in-progress book entitled 3D Game and Graphics Programming using UNIGINE. He is the CTO of BEiT Media.

Table of Contents

5
Using the AR User Framework

Section 3 – Building More AR Projects

6

Gallery: Building an AR App

7

Gallery: Editing Virtual Objects

8

Planets: Tracking Images

9

Selfies: Making Funny Faces

Other Books You May Enjoy

Index

Preface

Augmented Reality (**AR**) allows people to interact meaningfully with the real world through digitally enhanced content. This book will help get you started developing your own AR applications using the Unity 3D game engine and the AR Foundation toolkit provided by Unity Technologies Inc. Using the techniques and lessons presented in this book, you will be able to create your own AR applications and games for a variety of target devices.

AR technology is now commonly available on mobile consumer devices—smartphones and tablets, both iOS (ARKit) and Android (ARCore), and a new generation of wearable smart glasses. In this book, I focus on instructing you on how to develop AR applications for mobile devices, but the techniques and projects can also be applied to wearables.

By the end of this book, you will be able to build and run your own AR applications that add layers of information to the real world, enabling interaction with real and virtual objects and innovatively engaging your users.

Who this book is for

If you want to develop your own AR applications, I recommend using Unity with AR Foundation as it is one of the most powerful and flexible platforms. Developers interested in creating AR projects can use this book to accelerate their progress on the learning curve and gain experience through a variety of fun and interesting projects. This book complements Unity's own documentation and other resources and provides practical advice and best practices that will have you up and running and productive quickly.

You do not need to be a Unity expert to use this book, but some familiarity will help you get started more quickly. If you are a beginner, I recommend you first run through one or two introductory tutorials found on Unity Learn (https://learn.unity.com/). It may also be helpful to have some experience developing for mobile devices (iOS and/or Android).

That said, I start from the very beginning, walking you along your learning curve slowly at first and then faster as you gain experience. And I provide plenty of links to external resources if you want to learn more and explore specific topics in more depth. Experienced readers can push past the instructions and explanations they already know.

What this book covers

Chapter 1, Setting Up for AR Development, after briefly defining AR, gets you set up for AR development, installing the Unity 3D game engine and the AR Foundation toolkit, and ensuring your system is ready to develop for Android (ARCore) and/or iOS (ARKit) mobile devices.

Chapter 2, Your First AR Scene, jumps right into building and running AR scenes, starting with examples provided in the AR Foundation samples project from Unity, and then moving on to building your own simple scene from scratch, learning about ARSession components, prefabs, and a little bit of C# coding too.

Chapter 3, Improving the Developer Workflow, teaches you about troubleshooting, debugging, remote testing, and Unity MARS, which can make your development workflow more efficient.

Chapter 4, Creating an AR User Framework, sees you develop a framework for building AR applications that manages user interaction modes, user interface panels, and AR onboarding graphics, which we will save as a template for reuse in other projects in this book.

Chapter 5, Using the AR User Framework, is where you will build a simple AR place-on-plane application using the AR user framework created in the previous chapter, including a main menu and a PlaceObject mode and UI. This chapter also discusses some advanced issues, such as making AR optional, determining device support, and adding localization to your projects.

Chapter 6, Gallery: Building an AR App, is part one of a two-chapter project. Here, you will develop a picture gallery application that lets you hang virtual framed photos on your real-world walls. In the process, you will learn about UX design, managing data and objects, menu buttons, and prefabs.

Chapter 7, Gallery: Editing Virtual Objects, is the second part of the Gallery project, where you will learn to implement interactions with virtual objects in your AR scene, including selecting and highlighting, moving, resizing, deleting, collision detection, and changing the photo in your picture frame.

Chapter 8, Planets: Tracking Images, shows you how to build an educational AR app that uses image tracking of Solar System "planet cards" that instantiates virtual planets hovering and spinning above your table.

Chapter 9, Selfies: Making Funny Faces, is where you will learn to use the front-facing camera of your device to make fun and entertaining face filters, including 3D heads, face masks (with choice of material textures), and accessories such as sunglasses and mustaches. It also covers advanced features specific to ARCore and ARKit that may not be generally supported by AR Foundation itself.

To get the most out of this book

First, you need a PC or Mac capable of running Unity. The minimum requirements are not difficult; almost any PC or Mac today will be sufficient (see `https://docs.unity3d.com/Manual/system-requirements.html`).

If you are developing for iOS, you will need a Mac running OSX with the current version of XCode installed, and an Apple developer account. If you are developing for Android, you can use either a Windows PC or Mac.

It is not practical to develop for AR without a device capable of running your application. You should have an iOS device that supports Apple ARKit (search the web; Apple does not appear to publish a list – check here, for instance: `https://ioshacker.com/iphone/arkit-compatibility-list-iphone-ipad-ipod-touch`), or an Android device that supports ARCore (`https://developers.google.com/ar/discover/supported-devices`).

In *Chapter 1, Setting Up for AR Development,* I walk you through installing Unity Hub, the Unity Editor, XR plugins for your target device, the AR Foundation toolkit package, and other software to get you set up. The projects in this book are written and tested with Unity 2021.1.

As the technology is rapidly evolving, I try to focus on existing stable tools and techniques. Regarding software versions and installation instructions, naturally, things can change, and I recommend you use my instructions as guidelines but also look at online documentation (links usually given) for the most current instructions.

If you are using the digital version of this book, we advise you to type the code yourself or access the code from the book's GitHub repository (a link is available in the next section). Doing so will help you avoid any potential errors related to the copying and pasting of code.

Download the example code files

You can download the example code files for this book from GitHub at `https://github.com/PacktPublishing/Augmented-Reality-with-Unity-AR-Foundation`. If there's an update to the code, it will be updated in the GitHub repository.

We also have other code bundles from our rich catalog of books and videos available at `https://github.com/PacktPublishing/`. Check them out!

Download the color images

We also provide a PDF file that has color images of the screenshots and diagrams used in this book. You can download it here: `https://static.packt-cdn.com/downloads/9781838982591_ColorImages.pdf`.

Conventions used

There are a number of text conventions used throughout this book.

`Code in text`: Indicates code words in text, database table names, folder names, filenames, file extensions, pathnames, dummy URLs, user input, and Twitter handles. Here is an example: "Utilizing the Unity Input System package, we will first add a new `SelectObject` input action."

A block of code is set as follows:

```
public void SetPlacedPrefab(GameObject prefab)
{
    placedPrefab = prefab;
}
```

When we wish to draw your attention to a particular part of a code block, the relevant lines or items are set in bold:

```
using UnityEngine;

using UnityEngine.InputSystem;

public class GalleryMainMode : MonoBehaviour
{
    void OnEnable()
    {
```

```
        UIController.ShowUI("Main");
    }
}
```

Bold: Indicates a new term, an important word, or words that you see on screen. For instance, words in menus or dialog boxes appear in **bold**. Here is an example: "In the **New Scene** dialog box, select the **ARFramework** template."

Tips or important notes
Appear like this.

Get in touch

Feedback from our readers is always welcome.

General feedback: If you have questions about any aspect of this book, email us at customercare@packtpub.com and mention the book title in the subject of your message.

Errata: Although we have taken every care to ensure the accuracy of our content, mistakes do happen. If you have found a mistake in this book, we would be grateful if you would report this to us. Please visit www.packtpub.com/support/errata and fill in the form.

Piracy: If you come across any illegal copies of our works in any form on the internet, we would be grateful if you would provide us with the location address or website name. Please contact us at copyright@packt.com with a link to the material.

If you are interested in becoming an author: If there is a topic that you have expertise in and you are interested in either writing or contributing to a book, please visit authors.packtpub.com.

Section 1 – Getting Started with Augmented Reality

This section provides a basic introduction to developing AR applications and games with Unity, including AR technology concepts and how to use the Unity Editor. We cover the Unity XR plugin architecture, the AR Foundation toolkit, and other productivity tools. By the end of this section, you will be prepared to begin creating your own AR applications with Unity.

This section comprises the following chapters:

- *Chapter 1, Setting Up for AR Development*
- *Chapter 2, Your First AR Scene*
- *Chapter 3, Improving the Developer Workflow*

1
Setting Up for AR Development

Augmented reality (**AR**) is widely recognized as the next-generation computing platform where digital content is seamlessly merged into real-world experiences. This book will help get you started with developing your own AR applications using the Unity 3D game engine and the AR Foundation toolkit provided by Unity.

In this chapter, you will take your first steps by setting up your computer for AR development using the Unity 3D game engine. We will begin by briefly defining augmented reality, thus setting the context for this industry and some of the basics of AR technology. We will then install the Unity software, the AR Foundation toolkit, and make sure your system has been set up to develop for Android and/or iOS mobile devices. Finally, we'll build and run a test scene to verify things are working as they should.

We will cover the following topics:

- Defining augmented reality
- Getting started with Unity, including installation and using Unity
- Preparing your project for AR development, including XR plugins, AR Foundation, Input System, and the Universal Render Pipeline
- Setting up for mobile development (Android ARCore and iOS ARKit)

> **Note for Experienced Readers**
>
> If you are already familiar with Unity, already have it installed on your system, and are set up to build for your iOS or Android mobile device, you may be able to skim through details related to those topics that are interspersed in this chapter.

Technical requirements

First, you need a PC or Mac that's capable of running Unity. The minimum requirements are not difficult; almost any PC or Mac today will be sufficient (see `https://docs.unity3d.com/Manual/system-requirements.html`).

If you are developing for iOS, you will need a Mac running OSX with the current version of XCode installed, and an Apple developer account. If you are developing for Android, you can use either a Windows PC or Mac. We will discuss this further throughout this chapter.

It is not practical to develop for AR without a device capable of running your application. For this chapter (and this book as a whole), you will need either an iOS device that supports Apple *ARKit* (search the web as Apple does not appear to publish a list; for example, `https://ioshacker.com/iphone/arkit-compatibility-list-iphone-ipad-ipod-touch`) or an Android device that supports *ARCore* (`https://developers.google.com/ar/discover/supported-devices`).

Because this chapter is largely about installing tools and packages according to your requirements, please work through the topics in this chapter for additional technical requirements and to learn how to install them. The GitHub repository for this book can be found at `https://github.com/PacktPublishing/Augmented-Reality-with-Unity-AR-Foundation`.

Defining Augmented Reality

According to the Merriam-Webster dictionary, the word *augment* means "to make greater, more numerous, larger, or more intense," while *reality* is defined as "the quality or state of being real." Considering this, we realize that "augmented reality" is all about using digital content to improve our real world to add better information, understanding, and value to our experiences.

Augmented reality is most commonly associated with *visual* augmentation, where computer-generated graphics are combined with actual real-world visuals. When using a handheld mobile phone or tablet, for instance, AR combines graphics with the on-screen video (I call this *video see-through* AR). Using wearable AR glasses, graphics are directly added to your visual field (*optical see-through* AR).

But AR is not simply a computer graphic overlay. In his acclaimed 1997 research report, *A Survey of Augmented Reality* (`http://www.cs.unc.edu/~azuma/ARpresence.pdf`), Ronald Azuma proposed that AR must meet the following characteristics:

- *Combines the real and virtual*: The virtual objects are perceived as real-world objects that are sharing the physical space around you.

- *Interactive in real time*: AR is experienced in real time, not pre-recorded. For example, cinematic special effects that combine real action with computer graphics do not count as AR.

- *Registered in 3D*: The graphics must be registered to real-world 3D locations. For example, a **heads-up display** (**HUD**) where information is simply overlayed in the visual field is not AR.

To register a virtual object in 3D, the AR device must have the ability to track its location in 3D space and map the surrounding environment to place objects in the scene. There are multiple technologies and techniques for positional and orientation tracking (together referred to as *pose tracking*), as well as environmental feature detection, including the following:

- *Geolocation*: GPS provides low-resolution tracking of your location on the Earth (GPS accuracy is measured in feet or meters). This is usually good enough for wayfinding in a city and identifying nearby businesses, for example, but not for more specific positioning.

- *Image Tracking*: Images from the device's camera can be used to match the predefined or real-time 2D images, such as QR code markers, game cards, or product packaging, to display AR graphics that track an image's pose (3D position and orientation) relative to the camera space.

- *Motion Tracking*: Using the device's camera and other sensors (including inertial measurement by IMU motion sensors), you can compute your position and orientation in 3D, and detect visually distinct features in the environment. Academically, you may see this referred to as **Simultaneous Localization and Mapping** (**SLAM**).

- *Environmental Understanding*: As features are detected in the environment, such as X-Y-Z location depth points, they can be clusters to identify horizontal and vertical planes, as well as other shapes in 3D. These can be used by your application for object placement and interaction with real-world objects.

- *Face and Object Tracking*: Augmented selfie pictures use the camera to detect faces and map a 3D mesh that can be used to add a face mask or other (often humorous) enhancements to your image. Likewise, other shaped objects can be recognized and tracked, as may be required for industrial applications.

In this book, we will be using many of these techniques in real projects with Unity's AR Foundation toolkit, so that you can learn how to build a wide variety of AR applications. And we'll also be learning many other details and capabilities offered by Unity and AR software, all of which we'll use to improve the quality and realism of your graphics and provide engaging interactive experiences for your users.

Like all technologies, AR can potentially be used for better or for worse. A great exposé on a hypothetical disturbing future, where AR is ubiquitous and as consuming as today's mobile media technologies, can be found in this 2016 *Hyper-Reality* art video by Keiichi Matsuda (`http://hyper-reality.co/`). Hopefully, you can help build a better future!

Figure 1.1 – Hyper-Reality video by Keiichi Matsuda (used with permission)

In this book, we are using the Unity 3D game engine for development (`https://unity.com/`), as well as the AR Foundation toolkit package. AR Foundation provides a device-independent SDK on top of the device-specific system features provided by Google ARCore, Apple ARKit, Microsoft HoloLens, Magic Leap, and others. For further reading and to get a good introduction to mobile handheld augmented reality, check out the following links:

- ARCore Fundamental Concepts: `https://developers.google.com/ar/discover/concepts`.

- Introducing ARKit: `https://developer.apple.com/augmented-reality/arkit/`.

- Getting Started with AR Development in Unity: `https://developers.google.com/ar/discover/concepts`.

Let's start developing AR applications with Unity. First, you'll need to install Unity on your development computer.

Getting started with Unity

To develop AR applications with Unity, you need to install Unity on your development machine. In this section, we'll step through the installation process using Unity Hub, create a new Unity project, and introduce the basics of using the Unity Editor interface.

Installing Unity Hub

Unity Hub is a desktop application that serves as a portal to many of the resources developers may need to use Unity in their workflows. For now, we'll be using the **Installs** menu to install a version of the Unity Editor. Then, we'll use the **Projects** menu to create and manage our Unity projects. To do this, follow these steps:

1. Please download and install the **Unity Hub** program from `https://unity3d.com/get-unity/download`. Generally, you'll always want to use Unity Hub to install versions of Unity rather than downloading a Unity version installer directly.

2. If you haven't already, you may need to activate a Unity **User License**. This is free for the *Student* and *Community* plans; you can decide to upgrade to *Plus* or *Pro* at a later time. All license plans include the same versions of Unity; no features are disabled for free plans. The paid plans add access to professional cloud services that are very useful but not necessary for project development.

3.   Use the **Download Unity Hub** button, as shown in the following screenshot (you may need to agree to the *Terms of Service* first):

Figure 1.2 – Installing Unity Hub instead of downloading Unity directly

4.   With Unity Hub installed and open, you'll see menus for **Learn** and **Community**.

Clicking **Learn** takes you to *Unity Learn* projects and tutorials (including downloads for the various project assets). These can range from 5-minute tutorials to projects that take 15 hours to complete!

The **Community** menu provides links to many other Unity-hosted resources, including *Unity Now* conference talks, *Unity Blog, Answers, Q&As*, and *Forums*.

Now, let's install a version of the Unity Editor.

Installing a Unity Editor

When starting a new project, I like to use the latest **Official Release**, which has a leading edge without being a Beta or Alpha prerelease. If you are more cautious or have requirements to use the most stable release, choose the **long-term support** (**LTS**) version. These can be found under the **Recommended Release** heading in Unity Hub. Unity versions that are compatible with the writeups in this book are noted in the current .README file of this book's GitHub repository. Install a copy of the Unity Editor now, as follows:

1.   Select the **Installs** tab, then press **ADD** to open the **Add Unity Version** box.

2. From here, you can select a version of Unity to install.

> **Note – Unity Versions**
>
> In the current Unity version numbering system, the major release number (for example, Unity **2020**.x.x) loosely correlates with calendar years. The most stable versions are ones designated **LTS**, for **Long-Term Support**; for example, *Unity 2020.3.14f1 (LTS)*. LTS versions receive periodic maintenance and security updates but no new features. Point releases lower than LTS (for example, *Unity 2021.**1**.15f1*) are considered technical releases, which are reasonably stable while new features and bug fixes are currently in development. For the more adventurous, Beta and Alpha prereleases include cutting-edge features but with added risks.

3. Once you've selected the version of Unity you wish to install, click **Next** to see the **Add modules to your install** options. Here, you want to know what platforms and devices you expect to target with your projects.

    Module software can be quite large and may take time to install, so only pick what you know you will need soon. You can always come back later and add (or remove) modules as needed. Specifically, if you are developing your AR project for Android and ARCore, choose **Android Build Support**. If you are targeting iOS and ARKit, choose **iOS Build Support**. Likewise, if you are targeting other devices such as HoloLens or Magic Leap, choose the corresponding modules.

4. Depending on the modules you selected, you may need to press **Next** and accept an additional user license agreement. Then, press **Done** to download and install the software.

> **Tip – Where to Install Unity**
>
> Using the **gear** icon in the top-right of the Unity Hub window opens a **Preferences** window. Under the **General** preference tab, you can select the folder where your **User Editors** are installed on your computer. Since these can take up a considerable amount of disk space, you may not want to use the default location.

If you have any problems with Unity Hub or otherwise want to join in with discussions, visit the relevant section of the Unity community discussion forum at `https://forum.unity.com/forums/unity-hub.142/`.

Now, you're ready to create your first Unity project.

Creating and managing Unity projects

You will use Unity Hub to create new Unity projects. Projects are created in a specific folder on your system, with a set of subfolders populated with default settings and content based on the starting template you choose. Projects are opened with a specific version of Unity and continue to be associated with that specific version. To start a new project, complete the following steps:

1. Open the Unity Hub, select the **Projects** tab, and then click the **New** button. Notice the down arrow of the **New** button, which lets you select a Unity version to use for the new project that's different from the one you currently have installed.

2. The **Create New Project** box gives you the option to choose a **Project Name**, a **Location** where it should be created, and a starting **Template**. As shown in the following screenshot, I am selecting the **Universal Render Pipeline** template, in a folder named D:\Documens\UnityProjects:

Figure 1.3 – Creating a new project with the URP template in Unity Hub

> **Note – We're Using the Universal Render Pipeline**
>
> Unity offers multiple alternative render pipelines. The legacy "built-in" render pipeline sports better support from older third-party assets as it came before the newer **Scriptable Render Pipeline (SRP)** system (https://unity.com/srp), but the newer SRP-based pipelines are more performant and flexible. These include the **High Definition Render Pipeline (HDRP)** for high-quality rendering using high-end graphics hardware. There's also the **Universal Render Pipeline (URP)**, which is very fast, even on mobile devices, while providing excellent rendering quality. I recommend starting new AR projects with URP.

> **Tip – Avoid Spaces in Project Names**
>
> At the time of writing, there's a bug in some ARCore features that require that your project pathname contains no spaces, including the project name and all folder names up the tree.

3. After pressing **Create**, it may take a few moments for Unity to create your new project, import the default assets, and perform other setup steps before opening the Unity Editor window.

> **Tip – Upgrading Unity Projects**
>
> A great thing about Unity Hub is its ability to manage multiple versions of Unity and all your Unity projects. I tend to start new projects with the latest official release, though inevitably, new versions of the Unity Editor will be released. In general, it's best to stick with the version of Unity you used to create your project. If you need to upgrade to a newer version, do so cautiously and deliberately.
>
> Generally, going to a new minor update (for example, Unity 2021.2.**3** to 2021.2.**16**) is safe. Going to a point release (for example, Unity 2021.**2.x** to 2021.**3.x**) is usually OK but you may encounter unexpected problems. Upgrading to a new major release is an unusual event for me in my projects. In any of these cases, be sure that your project has been backed up (for example, on GitHub) before opening the project in a different version of Unity, and schedule time to resolve unforeseen problems.
>
> Unity includes automated tools to facilitate upgrading a project to a new version when it's opened in Unity. Your assets will be reimported. While upgrading to newer versions is supported, downgrading to a previous version is not.

When I create a new project in Unity, one of the first things I do is set **Target Platform** in **Build Settings** to the first platform that I know I will be using to develop and test my project. There are advantages to doing this as early as possible, as any new assets you add to the project will be imported and processed for your target platform. You are not *required* to do this now, but I do recommend that you perform the following steps. We will go into more detail later in this chapter (in the platform-specific topic sections).

With your project opened in Unity, follow these steps:

1. Open the **Build Settings** window by going to **File | Build Settings**.
2. In the **Platform** selection panel, choose your target platform. For example, if you're developing for Android ARCore, select **Android**, while if you're developing for Apple ARKit, choose **iOS**.

If the platform you require is not listed or is disabled, you may have forgotten to add the platform build module when you installed this version of Unity. Use **Unity Hub** to add the module now.

> **Tip – You Can Add Target Platform Modules via Unity Hub**
>
> If you are missing support for a target platform, open **Unity Hub**, click **Installs**, and then, for the specific Unity version you're using, click the 3-dot context menu and choose **Add Modules**. From there, you can use the checkboxes to add new modules.

3. You don't need to worry about the other build settings right now. Press the **Switch Platform** button. It may take a few minutes to reimport your project's assets.

At this point, your Unity Editor should have opened a new Unity project, showing a default URP **SampleScene**. Feel free to explore the editor windows and scene objects. It may look daunting at first, but we'll review the user interface next to help you get more comfortable.

Introducing the Unity Editor interface

When you open the Unity Editor for the first time, you will notice that it has a lot of separate window panels that contain different content. Let's explore these together.

The following screenshot shows **Unity Editor** with the Universal Render Pipeline template's **SampleScene**. The windows are arranged in a default layout. This "under construction" scene demonstrates many of the awesome rendering features of Unity that may or may not be relevant in an AR project. But let's focus on Unity itself for a moment:

Figure 1.4 – Unity Editor with the URP sample scene open

The Unity Editor is arranged in a layout of separate tabbed windows. An **Editor window** is a UI panel containing specific types of information and controls. More windows can be opened via the **Window** main menu. Let's review each window in the preceding screenshot and introduce some other fundamental terminology since you're getting to know Unity:

- **Hierarchy** window (*1*): The tree view of the current scene's GameObjects. Shows the same content as the scene in a hierarchical tree view of parent-child objects.

  You may have noticed in the preceding screenshot that, while examining both the **Scene** and **Hierarchy** windows, the **Safety Hat** GameObject is currently selected and highlighted. A Unity **GameObject** is an object that is part of a scene.

- **Scene** view window (*2*): This shows a 3D view of the current scene. Along the top of the scene window is an icon toolbar for controlling your working view of the scene.

- **Inspector** window (*3*): The components and properties of the selected GameObject.

  GameObjects have **components** attached that define the runtime behavior of a GameObject. Unity includes many built-in components, and you can write your own using the C# programming language. Each component may have individual **properties**; that is, settings that control the component.

  You can see that the Safety Hat has **Transform** and **Mesh Renderer** components, for instance.

  GameObjects always have one **Transform** component. GameObjects may also have a 3D mesh, renderer, and materials that determine how it's rendered in the scene. There are many other components you can add that extend an object's behavior, physics, and interactions.

- **Project** assets window (*4*): In this window, you'll find the files stored on the hard drive in your project's `Assets/` folder, located under the project's root directory.

  Assets include files that may be added to objects in a scene, such as images, audio, video, materials, and scripts. Scenes themselves are saved as assets. Complex predefined GameObjects can also be saved as assets, called **prefabs**.

- **Console** window (*4*, hidden behind Project tab): This shows error and information messages from your application.

- **Game** view window (*2*, hidden behind Scene tab): This shows the user's view, as rendered by the in-scene camera GameObject.

- **Main Menu**: At the top of the **Editor** window is a menu where you can access many features of Unity. Adding additional packages to your project may add more menu items.

- **Main Toolbar**: At the top of the **Editor** window and below the **Main** menu is an icon toolbar organized into three sections. On the left-hand side, there are tools for editing the Scene view (including **Move Tool**, **Rotate Tool**, and **Scale Tool**). In the center, there are the play mode controls (including **Play** and **Pause**). Finally, on the right, there are additional controls, including some that allow you to access your Unity account and cloud services.

Take a moment to explore the main menu items:

- The **File** menu is for creating, saving, and loading scenes and accessing your build settings.

- The **Edit** menu is for selecting and editing objects in the project, accessing project-specific settings and preferences, and other editor-related tools.

- The **Assets** menu provides tools for importing and managing project assets (as found in the **Project** window's `Assets/` folder).

- The **GameObject** menu lets you add new objects to the current scene.

- The **Component** menu provides a categorized list of components that you can add to the currently selected GameObject in the scene.

- The **Window** menu is where you can find and open additional windows that provide more features. Importing new packages into Unity may add new menu bar items.

> **Information – Using Play Mode in AR Projects**
>
> In most Unity projects, you can press the **Play** button (in the main toolbar) to go into *play mode* and run your scene in the Editor, running on your desktop rather than on the device. This is not so simple with an augmented reality scene since it requires an onboarding phase, where the software scans the environment for physical world features and then uses the physical device sensors for positional tracking. There are several solutions to facilitate your iterative developer workflow, all of which we will discuss in *Chapter 3, Improving the Developer Workflow*.

You can personalize and rearrange the editor's window layout to suit your needs and preferences. Layouts can be saved and loaded using the **Layout** selection menu in the top-right corner of the editor. The screenshots in this book use layouts that are different from Unity's default layout.

OK, enough talk – this is a hands-on book, so let's get hands-on right away and try out the Unity Editor.

Basics of using the Unity Editor

In this section, we'll build a trivial scene with a 3D cube that gives us more context to explain how to use Unity:

1. Create a new scene from the main menu by selecting **File | New Scene**.

2. A **New Scene** window will appear (Unity 2020+ only) that lets you select a scene template. Choose the one named **Basic (Built-in)**. Then, press **Create**.

    You will notice right away that the new scene contains two default GameObjects: a Main Camera and a Directional Light.

3. Add a 3D cube to the scene by clicking **GameObject | 3D Object | Cube**. With that, the Cube will be added to the scene and be visible in both the **Scene** and **Hierarchy** windows.

4. Ensure the Cube rests at the origin of our scene; that is, the (0, 0, 0) X-Y-Z coordinates. With **Cube** selected in the **Hierarchy** window, look in the **Inspector** window and set its **Transform | Position | X**, **Y**, and **Z** values to zero.

5. Let's rotate the Cube. In the same **Inspector Transform** component, set its **X-Rotation** value to -20.

    The scene may now look as follows:

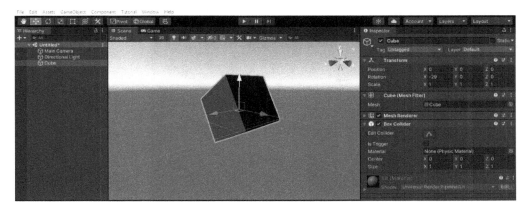

Figure 1.5 – New scene with a 3D cube

At this point, I encourage you to get familiar with the **Scene** view controls. Using a
3-button mouse on Windows, for example, *right-click* in the window to pivot the view,
Alt + left-click to orbit the view around the "center" of the view, and *center-click* the
mouse to move the view. To move closer or further out (zoom), use *Alt + right-click*
or use the scroll wheel. Note that the directional gizmo in the top right of the **Scene**
window indicates the current view showing the X, Y, and Z axes. For further information
(including one- or two-button mice), see https://docs.unity3d.com/Manual/
SceneViewNavigation.html.

> **Tip – RGB == XYZ**
>
> It's handy to remember that the red, green, and blue colors in gizmos
> correspond to the X, Y, and Z axes, respectively.

We modified the Cube's transform by editing its numerical values in the **Inspector**
window. You can also transform an object by directly manipulating it in the **Scene**
window. For example, in the main toolbar, select **Rotate Tool**. With the Cube currently
selected, you should now see the rotate gizmo rendered on the object in the scene. You can
grab one of the gizmo handles (X, Y, or Z) and drag it to rotate the object around that axis,
as shown in the following screenshot:

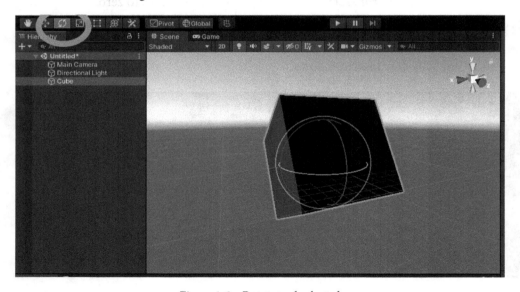

Figure 1.6 – Rotate tool selected

For more on transforming objects directly in the **Scene** window, see the Unity Manual
(https://docs.unity3d.com/Manual/PositioningGameObjects.html).

This was a very brief introduction to get you started. As a matter of habit, you should always save your work after accomplishing something. Let's save the scene, as follows:

1. From the main menu, select **File | Save As**, which will open the **Save Scene** window.

2. Navigate to the `Scenes/` subfolder (in your project's `Assets` folder).

3. Give the scene a name, such as `My Cube`, and press **Save**.

Tip – Confused or Overwhelmed? Take it a Step at a Time

As with any professional development and creative application, there's a huge assortment of things you can do with Unity, and it provides many tools to help you achieve your objectives. If you are confused or overwhelmed, a great strategy is to try and focus only on the menu items and windows you need right now and ignore the rest. We'll walk you through this with simple step-by-step instructions. As you gain experience and confidence, you'll expand your radius of familiarity and see how it all fits together. To be honest, I still learn new things about Unity each time I work on a project.

Of course, this was just a brief introduction to Unity. If you need to find out more, please head over to **Unity Learn**, where there are some excellent beginner tutorials (using the `https://unity.com/learn/get-started` link or the **Learn** tab in **Unity Hub**).

Also, take a look at the **Unity Manual** introductory topics (`https://docs.unity3d.com/Manual/UnityOverview.html`).

Organizing your project assets

You have access to your project assets in the **Project** window. I like to keep the project assets that I create in their own top-level folder, separate from other assets I might import from third-party sources such as the Unity Asset Store.

Likewise, Unity's URP project template includes **SampleScene** and example assets. I suggest moving the URP example assets into their own folder to keep them separate from your own application assets. You can do this by following these steps:

1. Create an `Assets` folder named `URP-examples`. In the **Project** window, click the + icon in the top left, select **Folder**, and name it `URP-examples`.

2. Drag each of the example folders into the `URP-examples` one, namely `ExampleAssets`, `Materials`, `Scenes`, `Scripts`, `TutorialInfo`, and the `Readme` file.

3. Leave the `Presets` and `Settings` folders in the root `Assets/` folder.

4. Create an `Assets` folder named `_App`. I like to prepend an underscore to this folder's name so that it remains at the top of the list.

5. Create child folders inside `_App/` named `Materials`, `Prefabs`, `Scenes`, and `Scripts`. These subfolders will remain empty for now, but we'll add to them as we work through this book.

Organizing your assets by file type is a common convention in Unity, but you may have your own way of doing things. Unity does not depend on these folder names or asset file locations. (That said, there are a few reserved folder names with special meanings to Unity; see `https://docs.unity3d.com/Manual/SpecialFolders.html`). Your **Project** window may now look as follows:

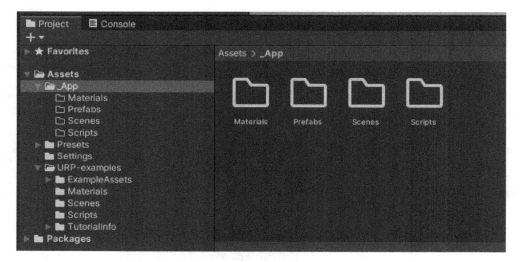

Figure 1.7 – Reorganized Project Assets folders

I think we're now ready to move on and continue setting up your system and installing the packages you need for AR development. We'll start by adding an AR device plugin to your project, and then do the same for the Foundation package.

Preparing your project for AR development

When you develop and build a project for augmented reality, Unity needs to know the device and platform you are targeting. This is a multi-step process that includes adding the device plugin to your project and setting the target platform in **Build Settings**. We'll address the device plugins now and **Build Settings** later in this chapter.

The following diagram shows the Unity XR technology architecture. As you can see, at the bottom of the stack are the various AR (and VR) provider plugins:

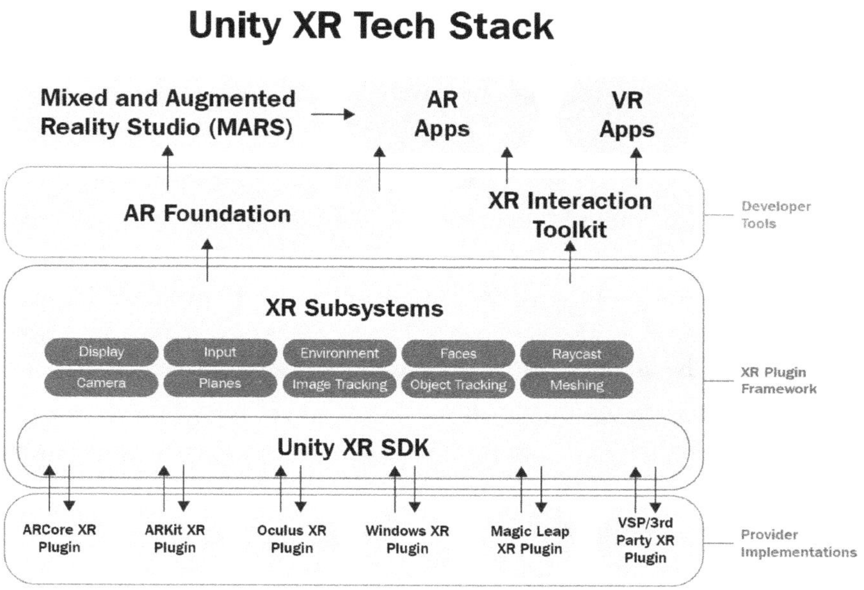

Unity XR Tech Stack

Figure 1.8 – The Unity XR tech stack

At the bottom of the stack is **XR Plugins**, separate provider packages that implement a software interface to a specific device. Plugins allow Unity to talk with a device by connecting the Unity XR subsystems with an operating system and runtime API. Ordinarily, you will not be using a plugin directly but a higher-level toolkit instead, such as AR Foundation (which we will install in the next section). Some plugins are provided and maintained by Unity Technologies; others are vendor-supported third-party plugins.

In the preceding diagram, at the top of the plugins are **XR Subsystems**, which form **XR Plugin Framework**. This abstracts sets of features into separate APIs. When an application is running, it can query the capabilities of the current runtime device and enable or disable sets of features in the app accordingly. Atop **XR Subsystems** is the **AR Foundation** toolkit (and **XR Interaction Toolkit**), which provides the main AR API for your Unity applications. We will be using AR Foundation extensively for the projects in this book.

Now, let's install the XR plugin(s) you need for this project.

Installing XR plugins for AR devices

To prepare our project for AR development, we'll install the AR device plugin for your target device via the **XR Plug-in Management** window. With your project open in Unity, follow these steps:

1. Open the **Project Settings** window by selecting **Edit | Project Settings** from the main menu.

2. In the **Settings** menu on the left, select **XR Plugin Management**.

3. Click the **Install XR Plugin Management** button. It may take a moment for Unity to import and compile the package scripts.

4. If necessary, click the **XR Plug-in Management** item again to show **Plug-in Providers** and other options. Notice that there are tabs for each of the target platforms. Select the one you will be targeting first.

    For example, in the **XR Plug-in Management** window, the **Android** tab will be only available if you installed the **Android Build Support** module when you installed Unity via Unity Hub.

5. Check the checkbox for the AR plugin you want to use. For example, for Android, select **ARCore**, while for iOS, select **ARKit**.

Tip – Don't Mix VR and AR Plugins in the Same Project

You'll see that the **XR Plug-in Management** window lets you choose any combination of AR and VR plugins. In our projects, we're only interested in the AR ones. Generally, do not include both AR and VR plugins in the same project as the build settings, player settings, camera rigs, and many other things can differ significantly between AR and VR projects. (Perhaps when you read this, there will be devices that support both modes in a single app, but I am not aware of any at this time.)

In the following screenshot of the **Project Settings** window, I have selected the **XR Plug-in Management Settings** menu. In my window, there are three tabs for each of the possible target platforms for this project that I have installed: Desktop, iOS, and Android (yours may be different). With the **Android** tab selected, you can see that I have checked the **ARCore** plugin. You'll also notice that, on the left-hand side, there's an additional **ARCore** menu item that you can click to see options that are specific to that plugin:

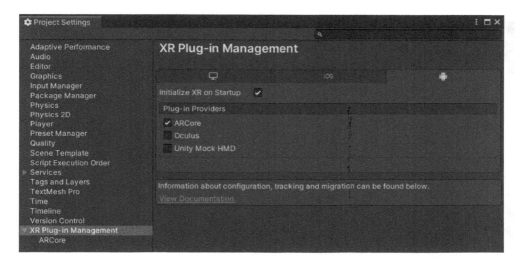

Figure 1.9 – XR Plug-in Management window with the ARCore plugin selected

Interestingly, **XR Plug-in Manager** is a shortcut to installing the corresponding packages in **Package Manager**. You can verify this by opening **Package Manager** and reviewing the installed packages by performing the following steps:

1. Open Package Manager from the main menu and choose **Window | Package Manager**.

2. Ensure the filter selection at the top left of the **Package Manager** window says **Packages In Project**.

3. You should see your plugin in the list; for example, **ARCore XR Plugin**.

For example, in the following screenshot of **Package Manager**, which shows **Packages In Project** (top left of the window), **ARCore XR Plugin** has been installed and selected. You can see that this specific version of the plugin has been **Verified** for the Unity version being used by this project. It also shows a description of the plugin's features, links to its documentation, and other details. Also, I have unfolded the plugin's **Other Versions** list to show you how to review each of the plugin's versions; this is where you might upgrade (or downgrade) a plugin to a different version:

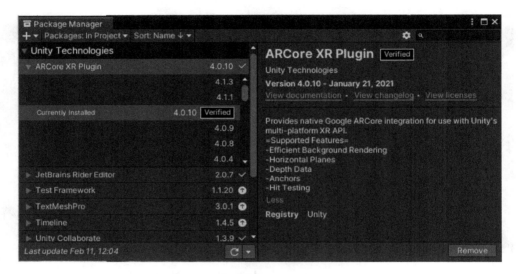

Figure 1.10 – Package Manager with the ARCore XR plugin installed in this project

At this point, you could begin developing an augmented reality project, if you wanted to write code directly using the XR subsystem's developer-facing C# interface. However, it's more likely that you will install a higher-level toolkit that is more Unity developer-friendly. Still, you may need to drop down into the plugin framework to access the XR subsystems directly. For example, you may wish to scan and start a particular subsystem, as shown in the example at `https://docs.unity3d.com/Manual/xrsdk-runtime-discovery.html`. Later in this book, we may need to access the plugin framework's SDK. For the most part, we will be using the higher-level **AR Foundation** toolkit.

Installing the AR Foundation package

AR Foundation is a package that provides a development layer between your application and the underlying device features and plugins. AR Foundation provides components and other assets that help you build AR projects once, then deploy for multiple mobile and wearable AR devices. Using a "unified workflow," as Unity says, your app can support current and future features that may or may not be currently available on your end user's specific device at runtime. This helps "future-proof" your AR apps. In this section, we'll install and explore AR Foundation.

The features that are supported by AR Foundation will depend on the current capabilities of the target devices and varies between versions of AR Foundation. The following chart shows the feature support per platform that AR Foundation offers:

AR Foundation Feature	ARCore	ARKit	Magic Leap	HoloLens
Device tracking	yes	yes	yes	yes
Plane tracking	yes	yes	yes	
Point clouds	yes	yes		
Anchors	yes	yes	yes	yes
Light estimation	yes	yes		
Environment probes	yes	yes		
Face tracking	yes	yes		
2D Image tracking	yes	yes	yes	
3D Object tracking		yes		
Meshing		yes	yes	yes
2D & 3D body tracking		yes		
Collaborative participants		yes		
Human segmentation		yes		
Raycast	yes	yes	yes	
Pass-through video	yes	yes		
Session management	yes	yes	yes	yes
Occlusion	yes	yes		

Figure 1.11 – AR Foundation 4.1.5 features per platform

See the **Platform Support** section of the AR Foundation documentation page (`https://docs.unity3d.com/Packages/com.unity.xr.arfoundation@latest/index.html`) for the most up to date details for the version you are using.

Unity provides a **Package Manager**, which enables you to expand Unity's core functionality by installing additional packages in your project. This way, you can choose just the features you need for a particular project. The XR plugins we installed in the previous section are packages. Now, we'll use the Package Manager to install the **AR Foundation** package.

> **Information – Advantages of Unity Packages**
>
> With Unity packages come many advantages. Unity can update the core Editor independently of other features. Likewise, packages can be updated outside of Unity's core release cycle. Decoupling their dependencies reduces the risk of schedule delays and technical problems, allowing for more agile development cycles and support for technology advances inside and outside the Unity offices. For example, if Apple releases an update to ARKit, then Unity can release an update to its ARKit plugin without having to wait for the next release of the Unity Editor, nor depending on the Unity core development team. If you've ever worked on a large project with multiple teams, you can appreciate the benefits of this architecture. Teams can be organized so that they focus on the details that their package provides, and then test for successful integration with the Unity core product.

You can install AR Foundation using Package Manager by following these steps:

1. Open the Package Manager from the main menu by going to **Window | Package Manager**.
2. Set the package filter in the top left to **Unity Registry** to see a list of all the official packages.
3. In the search box, type `ar`. You should now see all the AR-related packages in the list.
4. It is important to pay attention to the version numbers of the package, and whether that particular version has been verified with the Unity version you're using in your project.
5. Select **AR Foundation**, and then press **Install**. It may take a moment to install.

Once installed, you may discover that new items have been added to the main menu bar, including options under **GameObject | XR**. Don't select any just yet – we'll get to that in the next chapter, *Chapter 2, Your First AR Scene*, where we will use the toolkit to create our first AR scene with AR Foundation.

You also need to choose an input handler for your project. We'll look at this in the next section.

Choosing an input handler

The Unity product is continually improving. One relatively recent advancement is the introduction of the new **Input System**, which is replacing the classic **Input Manager**. At the time of writing, Unity projects can be configured to use either one, or both in the same project. The input handler you choose can have a significant impact on your development because their usages are quite different. The classic Input Manager mostly uses *polling*, while the new Input System uses *events* (see `https://blog.unity.com/ technology/introducing-the-new-input-system`). This is a generalization as both software patterns can be implemented using either handler, but the new Input System is better designed and more flexible. In the interest of advancing the state of the art, the projects in this book will use the new Input System.

However, some example scenes that you will be importing into your project, including the *AR Foundation Samples* in *Chapter 2, Your First AR Scene*, will use the classic Input Manager, so it's prudent to allow your project to support both.

To configure your project to use the new Input System, perform the following steps:

1. To import the Input System package, open the Packager Manager by going to **Window | Package Manager**.

2. Select **Unity Registry** from the filter selection in the top left of the window.

3. Find **Input System** (use the search field and type in `input`), and click **Install**.

4. You may be prompted to let Unity automatically change your Player Settings to use the new Input System. You can say "no" to this. We'll do this manually.

5. Open the **Player Settings** window by going to **Edit | Project Settings | Player**.

6. Locate **Configuration | Active Input Handling** and select **Both** (or if you prefer, select **Input System Package (New)**).

We will begin working with input in *Chapter 2, Your First AR Scene*, as well as the subsequent chapters.

You also need to set up the project's render pipeline for AR support. Let's learn how to do this.

Adding support for the Universal Render Pipeline

Because we created this project using the **Universal Render Pipeline** (**URP**), there's one additional thing you need to do – add AR video background support to the graphics forward renderer (see `https://docs.unity3d.com/Packages/com.unity.xr.arfoundation@4.1/manual/ar-camera-background-with-scriptable-render-pipeline.html`). This feature renders the device's video feed immediately on the screen before the virtual graphics are rendered on top of those pixels. Perform the following steps:

1. In the **Project** window, locate the folder that contains the Scriptable Render Pipeline settings assets. This is usually the `Assets/Settings/` folder.

2. Select the asset named **ForwardRenderer**.

3. In the **Inspector** window, click the **Add Renderer Feature** button and select **AR Background Renderer Feature**. The resulting Forward Renderer settings are shown in the following screenshot:

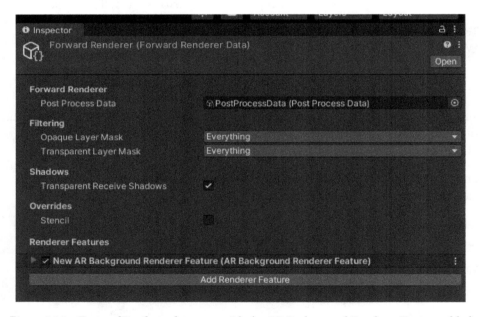

Figure 1.12 – ForwardRenderer data asset with the AR Background Renderer Feature added

Furthermore, as a reminder, if you import any assets into your project, you may need to convert their materials for the render pipeline. We will do this for the sample assets at the end of this chapter.

You have now installed Unity using Unity Hub, created and opened a new Unity project, installed XR plugins for your AR device, installed the AR Foundation package, selected an input handler for your project, and configured the render pipeline for AR. The next step is to continue setting up your project for the target platform.

Setting up for mobile development

Depending on the device platform you are targeting with your project, you will need to install additional software and development tools, as well as configure your Unity project with platform-specific settings.

If you are developing for Android ARCore, go to the *Setting up for Android/ARCore Development* section. If you are developing for Apple ARKit, go to the *Setting Up for iOS/ARKit Development* section. Finally, if you are developing for wearable AR devices, go to the *Developing for wearable AR glasses* section.

Setting up for Android/ARCore development

If you want to build and run your project on an Android device, there are a few extra steps in setting up your project for Android development and ARCore. I'll summarize the process here but naturally, things can change, and I recommend that you look at the documentation for the most current instructions, including Google's ARCore docs, in addition to the Unity Manual. Here is a list of some relevant links:

- ARCore Supported Devices: `https://developers.google.com/ar/devices`

- Unity Manual – Android Environment Setup: `https://docs.unity3d.com/Manual/android-sdksetup.html`

- ARCore Unity – Overview of Features: `https://developers.google.com/ar/develop/unity`

- Unity ARCore Extensions Installation: `https://developers.google.com/ar/develop/unity-arf/enable-arcore`

- Unity ARCore Plugin: `https://docs.unity3d.com/Packages/com.unity.xr.arcore@4.1/manual/index.html` (find the doc page for the version you are using)

You may have already completed the first few steps for setting up for Android and ARCore development, but I'll repeat them here briefly:

1. *Build support modules*: In Unity Hub, ensure you have installed the **Android** platform build support modules for the specific version of Unity you are using with your project.

    In the Unity Hub **Add Modules** window, there's a > icon for unfolding **Android Build Support**. Ensure you have checked the checkboxes for both **Android SDK & NDK Tools** and **OpenJDK**.

    Note that if you need to customize the location of either the Android SDK, NDK, or JDK libraries, use the **Unity Preferences** window in the Unity Editor, by navigating to **Edit | Preferences | External Tools**, and specify the paths for Unity to find where you installed these separately.

2. *Target Platform*: In the Unity Editor, open the **Build Settings** window by selecting **File | Build Settings**. Within the **Platforms** panel, select the **Android** platform from the list. If it is not already selected, please click the **Change Platform** button. If it is disabled, go back to *Step 1*.

3. *XR Plugin*: Ensure the **ARCore** plugin is currently installed and selected. Select **Edit | Project Settings**, and then select **XR Plug-in Management** from the side menu (initializing it if necessary). Click the **Android** icon to see the list of Android plugins and check the **ARCore** checkbox if it is not already checked.

4. *USB Debugging*: The next step is to enable USB debugging on your Android device (phone or tablet). Open the device's **Settings > About** window and find the **Build Number** item. (Depending on the brand, you may need to drill down another level or find the **Build Number** item in a slightly different location.) The next thing you must do I think is very funny – perform a magical incantation by clicking the **Build Number** item seven times! Then, magically, a **Developer Options** menu option will appear. Select that and enable **USB Debugging**.

    You can now connect your device to your development machine, and it should be recognized as an attached peripheral device.

The next thing to consider is the **Android Player** settings in your project. A review of these options can be found here: `https://docs.unity3d.com/Manual/class-PlayerSettingsAndroid.html`. Specific settings are required for AR projects targeting ARCore. Please double-check the current requirements, as can be found in the *Configure Project Settings* topic on the *Quickstart ARCore* page (`https://developers.google.com/ar/develop/unity-arf/quickstart-android`). Continuing from the previous steps, I suggest doing the following:

1. *Player Settings*: In Unity, navigate to **Edit | Project Settings | Player** to open the **Player Settings** window. It contains many options, including tabs at the top to switch between platform-specific settings. Generally, you can keep the default settings unless otherwise advised, or when you're optimizing your project builds. Initialize the following settings:

2. **Other Settings | Rendering**: Uncheck **Auto Graphics API**. If Vulkan is listed under Graphics APIs, remove it as Vulkan is not yet supported by ARCore. To do so, select **Vulcan** and press the - (minus) icon in the lower right. Also, uncheck **Multithreaded Rendering** as it's (currently) not compatible with ARCore.

3. **Other Settings | Package Name**: Create a unique app ID using a Java package name format. Unity chooses a default based on your project name; for example, `com.DefaultCompany.MyARProject`.

4. **Other Settings | Minimum API Level**: If you are building an **AR Required** app, specify **Android 7.0 'Nougat' (API Level 24)** or higher. If you are building an **AR Optional** app, specify Android **API Level 14** or higher.

Information – The Word "player" in Unity

The word "*player*" in Unity carries multiple meanings. The user of your application or game may be referred to as the *player*. In a game, the first-person GameObject (containing a camera controlled by the user) might also be referred to as the *player*. In a non-AR video game, the game controller might be called the *player controller*. However, in **Project Settings**, the *player* refers to the result of the build process; it is an executable program that is installed on your target device (along with other asset files and data) that "*plays*" your application. In this case, the word is akin to a *media player*, for example, that plays a music or video file. **Player Settings** in Unity configures how Unity is built and deployed to your target device.

Meanwhile, you also have the option to install additional capabilities provided by the **ARCore Extensions** package for Unity. This package extends AR Foundation to some more advanced features of ARCore that are currently not supported in AR Foundation. To install **ARCore Extensions**, perform the following steps:

1. Download the latest `arcore-unity-extensions-*.tgz` tarball from the GitHub releases page at `https://github.com/google-ar/arcore-unity-extensions/releases/`.

2. Open the Package Manager using **Window | Package Manager**.

3. In the top left of the window, click the + icon and choose **Add package from tarball**, as shown here:

Figure 1.13 – Adding a tarball package

4. Locate the downloaded `arcore-unity-extensions-*.tgz` tarball.

5. Then, click **Open**. It may take a few moments to install the package and any dependencies.

Your project is now set up to target Android ARCore with AR Foundation. We'll verify your settings in the next chapter, *Chapter 2, Your First AR Scene*, when we create an AR scene, build it, and run it on your device.

Setting up for iOS/ARKit development

If you want to build and run your project on an Apple iOS device, there are a few extra steps in setting up your project for iOS development and ARKit. I'll summarize the process here, but naturally, things can change, and I recommend that you look at the necessary documentation for the most current instructions.

Developing for iOS requires a Mac computer running OSX. Then, you need to install the XCode development environment. It is also strongly recommended that you join the Apple Developer Program, which currently costs $99 (USD) per year for individuals. You can do some limited Unity development for iOS without becoming an Apple Developer but it's not practical, especially for AR, where you need to test your app on a physical device.

Here is a list of some relevant links:

- Apple Developer Program: `https://developer.apple.com/programs/`

- Unity Manual – Getting Started with iOS Development: `https://docs.unity3d.com/Manual/iphone-GettingStarted.html`

- Unity Manual – Building for iOS: `https://docs.unity3d.com/Manual/UnityCloudBuildiOS.html`

- Unity ARKit Plugin: `https://docs.unity3d.com/Packages/com.unity.xr.arkit@4.1/manual/index.html` (find the doc page for the version you are using)

Information – How to Develop for iOS Without a Mac

While iOS development requires a Mac computer running OSX, it's possible to work around this using **Unity Cloud Builds**. This process is not for beginners, nor those timid about DevOps procedures. You will still need access to a Mac development machine to set up your Apple license, provisioning profile, iOS certificate, and p12 file, but then you can use those to set up a Unity Cloud Build for iOS. See `https://docs.unity3d.com/Manual/UnityCloudBuildiOS.html` for more information. After each successful build, you'll download the built application's `.ipa` file to your iOS device. This does not lend itself to a rapid development cycle! If you're in this situation, my recommendation is to buy a used Android phone that supports ARCore. Then, develop your app using AR Foundation on your Windows PC targeting Android first, and then periodically run iOS/ARKit builds to test and verify it runs on that device. Unity Cloud Builds requires a Unity Plus or Pro license or a Unity Teams Advanced subscription.

Developing for iOS and ARKit requires performing the following steps. You may have completed some of these steps already:

1. *Apple Developer Program*: This is your admission ticket for developing for iOS. Go to `https://developer.apple.com/programs/` to learn more and enroll.

2. *Xcode*: Download and install the current copy of Xcode, the development environment required to develop any Apple products. It's available on the Mac App Store: `https://apps.apple.com/us/app/xcode/id497799835`.

3. *Build support modules*: In **Unity Hub**, ensure you have installed the **iOS** platform build support modules for the specific version of Unity you are using with your project.

4. *Target Platform*: In the Unity Editor, open the **Build Settings** window by selecting **File | Build Settings**. Within the **Platforms** panel, select the **iOS** platform from the list. If it is not already selected, please click the **Change Platform** button. If it is disabled, go back to *Step 1*.

5. *XR Plugin*: Ensure the **ARKit** plugin is currently installed and selected. Select **Edit | Project Settings**, and then select **XR Plug-in Management** from the side menu (initialize it if necessary). Click the **iOS** tab to see the list of iOS plugins, and check the **ARKit** checkbox if it is not already checked.

6. *Player Settings*: In the **Edit | Project Settings | Player Settings** window, there are settings you may need to use, including checking the **Requires ARKit** checkbox, providing a text value for **Camera Usage Description** (such as `Required for augmented reality support`), setting **Target minimum iOS Version** to `11`, and **Architecture | ARM64**.

When Unity builds an iOS project, it does not actually build the app. Instead, it constructs an XCode project folder that is then opened in XCode, which, in turn, is used to build the app. One of the critical services XCode provides is ensuring you are authorized for development by provisioning your app, including the following:

- Installing a **Development Provisioning Profile** for each device where you plan to test your app. Follow the instructions at `https://docs.unity3d.com/Manual/UnityCloudBuildiOS.html`, under the *Create a Certificate* topic.

- Adding your **Apple ID** account to Xcode by going to **Preferences | Accounts**.

For more information on using Xcode and Unity, see the *Unity Manual: Structure of a Unity Xcode Project* (`https://docs.unity3d.com/Manual/StructureOfXcodeProject.html`) and other related pages.

This process can be confusing. Everyone who develops for iOS goes through a similar process, so you're certainly not alone, and there's a lot of answers to be found on the internet. Remember: "DuckDuckGo is your friend." And fortunately, you usually only need to do this once.

Note that you can also set your **Signing Team ID** in your Unity Player settings by navigating to **Edit | Project Settings | Player | Identification**.

> **Information – Apple's Own AR Development Tools**
>
> Upon reviewing the Apple web pages, you will discover that they provide their own AR development tools (`https://developer.apple.com/augmented-reality/tools/`) apart from Unity. Of course, I'm a big fan of Unity and AR Foundation, which give you device independence and all the other powerful features of Unity, but it's good to be aware of alternatives.

Your project has now been set up to target Apple ARKit with AR Foundation. We'll verify your settings in the next chapter, *Chapter 2, Your First AR Scene*, when we create an AR scene, build it, and run it on your device.

Developing for wearable AR glasses

AR Foundation supports not just handheld mobile AR devices using ARCore and ARKit, but also wearable AR glasses, including Microsoft HoloLens and Magic Leap. Likewise, targeting wearable AR devices may require configuring Unity to target a platform other than Android or iOS. Wearable AR glasses remain relatively expensive and outside the reach of the typical consumer as they're aimed at corporate or industrial applications. While this book can serve as a lovely starting point for developing these devices, and the projects can be adapted accordingly, it is outside the scope of this book to support wearable AR devices in the subsequent chapters.

For **Microsoft HoloLens**, you must set up Unity to target **Universal Windows Platform (UWP)**, beginning with installing the required module via Unity Hub, as shown in the following screenshot:

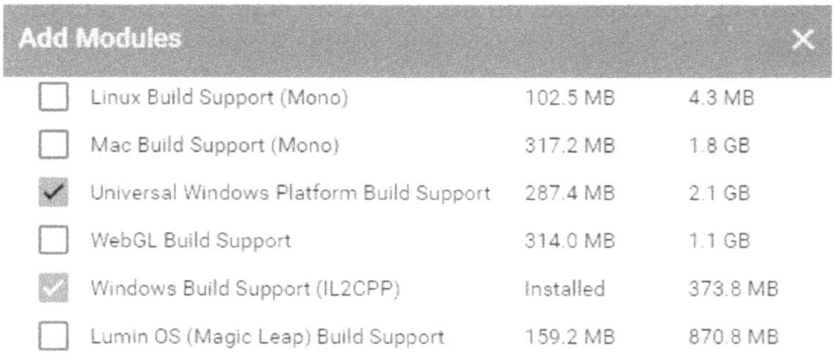

Figure 1.14 – Adding UWP build support for HoloLens

To set up for HoloLens development, you will need to use **Visual Studio IDE** and a compatible version of **Windows 10 SDK**. For additional information, here are some useful links:

- Unity for Windows Mixed Reality: `https://unity3d.com/partners/microsoft/mixed-reality`.

- Microsoft Mixed Reality – Install the Tools: `https://docs.microsoft.com/en-us/windows/mixed-reality/develop/install-the-tools?tabs=unity` (this includes an installation checklist too).

- Unity Windows XR Plugin: `https://docs.unity3d.com/Packages/com.unity.xr.windowsmr@5.2/manual/index.html`. You just have to find the document page for the version you are using. This page also includes recommended **Build Settings** and **Player Settings**.

> **Information – Microsoft Mixed Reality Toolkit (MRTK)**
>
> Note that Microsoft also offers its own open source cross-platform development kit, known as the **Mixed Reality Toolkit** (**MRTK**), for Unity, an alternative to AR Foundation. I think this framework has a very interesting implementation with a versatile architecture that supports a spectrum of devices from AR to VR. Learn more here: `https://docs.microsoft.com/en-us/windows/mixed-reality/develop/unity/mrtk-getting-started`.

For the **Magic Leap** wearable AR products, you must set up Unity to target **Lumen OS**, beginning with installing the required module via Unity Hub, as shown in the following screenshot:

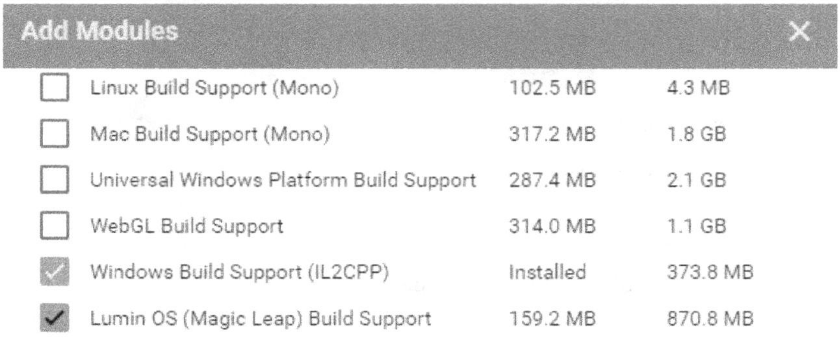

Add Modules			×
☐ Linux Build Support (Mono)	102.5 MB	4.3 MB	
☐ Mac Build Support (Mono)	317.2 MB	1.8 GB	
☐ Universal Windows Platform Build Support	287.4 MB	2.1 GB	
☐ WebGL Build Support	314.0 MB	1.1 GB	
☑ Windows Build Support (IL2CPP)	Installed	373.8 MB	
☑ Lumin OS (Magic Leap) Build Support	159.2 MB	870.8 MB	

Figure 1.15 – Adding Lumen OS build support for Magic Leap

For additional information, here are some useful links:

- Unity for Magic Leap: `https://unity3d.com/partners/magicleap`
- Magic Leap Developer Portal: `https://developer.magicleap.com/en-us/home`
- Magic Leap Unity Development: `https://developer.magicleap.com/en-us/learn/guides/unity-overview`
- Using Magic Leap with AR Foundation: `https://resources.unity.com/unitenow/onlinesessions/using-magic-leap-with-ar-foundation-in-unity-2020-1` (*Unite Now* presentation)

Interestingly, Magic Leap provides a **Unity Template** that you can add to Unity Hub as a starting point for new projects (`https://github.com/magicleap/UnityTemplate`).

Now that you have a project set up for AR development on your target platform and device, let's build a test to make sure things are working so far.

Building and running a test scene

Before moving on and building an AR project, it is prudent to verify your project has been set up properly so far by trying to build and run it on your target device. For this, we'll create a minimal AR scene and verify that it satisfies the following checklist:

- You can build the project for your target platform.
- The app launches on your target device.
- When the app starts, you see a video feed from its camera on the screen.
- The app scans the room and renders depth points on your screen.

I'll walk you through this step by step. Don't worry if you don't understand everything; we will go through this in more detail together in *Chapter 2, Your First AR Scene*. Please do the following in your current project, which should be open in Unity:

1. Create a new scene named *BasicTest* by selecting **File | New Scene**, then **Basic (Built-In)** template, then **File | Save As**. From here, navigate to your `Scenes` folder, call it `BasicTest`, and click **Save**.
2. In the **Hierarchy** window, delete the default **Main Camera** (*right-click* and select **Delete**, or use the *Del* keyboard key).
3. Add an AR Session object by selecting **GameObject | XR | AR Session**.

4. Add an AR Session Origin object by selecting **GameObject | XR | AR Session Origin**.

5. Add a point cloud manager to the Session Origin object by clicking **Add Component** in the **Inspector** window. Then, enter `ar point` in the search field and select **AR Point Cloud Manager**.

You will notice that the Point Cloud Manager has an empty slot for a Point Cloud Prefab, which is used for visualizing the detected depth points. A **prefab** is a GameObject saved as a project asset that can be added to the scene (*instantiated*) at runtime. We'll create a prefab using a very simple Particle System. Again, if this is new to you, don't worry about it – just follow along:

1. Create a Particle System by selecting **GameObject | Effects | Particle System**.

2. In the **Inspector** window, rename it `PointParticle`.

3. On the **Particle System** component, uncheck the **Looping** checkbox.

4. Set its **Start Size** to `0.1`.

5. Uncheck the **Play on Awake** checkbox.

6. Click **Add Component**, enter `ar point` in the search field, and select **AR Point Cloud**.

7. Likewise, click **Add Component** and select **AR Point Cloud Visualizer**.

8. Drag the **PointParticle** object from the **Hierarchy** window to the **Prefabs** folder in the **Project** window (create the folder first if necessary). This makes the GameObject into a prefab.

9. Delete the **PointParticle** object from the **Hierarchy** window using *right-click | **Delete** or press the *Del* key.

The Inspector window of the **PointParticle** object should now look as follows:

Figure 1.16 – Inspector view of our PointParticle prefab with the settings we're using highlighted

We can now apply the PointParticle prefab to the AR Point Cloud Manager, as follows:

1. In the **Hierarchy** window, select the **AR Session Origin** object.

2. From the **Project** window, drag the **PointParticle** prefab into the **AR Point Cloud Manager | Point Cloud Prefab** slot. (Alternatively, click the "doughnut" icon to the right of the slot to open the **Select GameObject** window, select the **Assets** tab, and choose **PointParticle**).

3. Save the scene using **File | Save**.

The resulting AR Session Origin should look as follows:

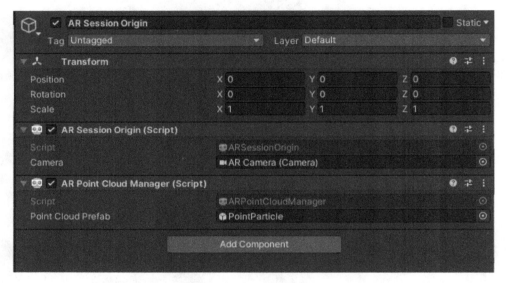

Figure 1.17 – Session Origin with a Point Cloud Manager component populated with the PointParticle prefab

Now, we are ready to build and run the scene. Perform the following steps:

1. Open the **Build Settings** window using **File | Build Settings**.

2. Click the **Add Open Scenes** button to add this scene to the build list.

3. In the **Scenes in Build** list, uncheck all scenes except the **BasicTest** one.

4. Ensure your device is connected to your computer via USB cable.

5. Press the **Build And Run** button to build the project and install it on your device. It will prompt you for a save location; I like to create a folder in my project root named `Builds/`. Give it a filename (if required) and press **Save**. It may take a while to complete this task.

If all goes well, the project will build, install on your device, and launch. You should see a camera video feed on your device's screen. Move the phone slowly in different directions. As it scans the environment, feature points will be detected and rendered on the screen. The following screen capture shows my office door with a point cloud rendered on my phone. As you scan, the particles in the environment that are closer to the camera appear larger than the ones further away, contributing to the user's perception of depth in the scene.

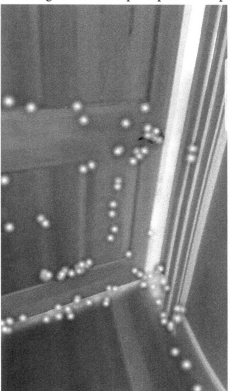

Figure 1.18 – Point cloud rendered on my phone using the BasicTest scene

If you encounter errors while building the project, look at the **Console** window in the Unity Editor for messages (in the default layout, it's a tab behind the **Project** window). Read the messages carefully, generally starting from the top. If that doesn't help, then review each of the steps detailed in this chapter. If the fix is still not apparent, do an internet search for the message's text, as you can be certain you're probably not the first person to have a similar question!

> **Tip – Build Early and Build Often**
>
> It is important to get builds working as soon as possible in a project. If not now, then certainly do so before the end of the next chapter, as it does not make a lot of sense to be developing an AR application without having the confidence to build, run, and test it on a physical device.

With a successful build, you're now ready to build your own AR projects. Congratulations!

Summary

In this chapter, after a quick introduction to augmented reality, you immediately got started on your road to developing your own AR projects. You installed Unity via Unity Hub and learned the importance of tracking the different versions of Unity, as well as its projects and packages. You got a brief tour of using the Unity Editor, including some key concepts that are fundamental to 3D and AR.

You then set up your project and system software for AR development, including installing an XR plugin, the AR Foundation package, tools for Android or Xcode, and other items necessary to get things set up. Lastly, we created a minimal AR scene (including a quick point cloud prefab using a particle system component) and built the scene to verify it builds and runs on your target device.

Setting up your machine may be intricate and painful, but it's your entry ticket to Unity development, and everyone has to do it. If you got through this chapter with everything running, you are a hero!

In the next chapter, we'll begin to take a closer look at AR development using Unity and AR Foundation by creating a new AR scene, step by step, explaining each component as we go.

2
Your First AR Scene

Creating a simple Augmented Reality (AR) scene is quite simple with Unity AR Foundation. The steps involved might only take a page or two. However, when we create a scene together in this chapter, each step will be explained in context so that you can gain a full understanding of what comprises an AR scene using AR Foundation.

But before we do that, we'll take a look at some AR examples provided by Unity, including the AR Foundation Samples project, and build their example scenes for your device. And because that project contains some useful assets, we'll export those as an asset package for reuse in our own projects.

In this chapter, we will cover the following topics:

- Building and running the AR Foundation Samples project
- Exporting and importing sample assets
- Constructing a new Unity AR scene
- Introduction to C# programming and the MonoBehaviour class
- Using AR raycast to place an object on a plane
- Instantiating a GameObject
- Creating and editing prefabs

Technical requirements

To implement the project provided in this chapter, you will need Unity installed on your development computer, connected to a mobile device that supports augmented reality applications (see *Chapter 1, Setting Up for AR Development*, for instructions). The completed project can be found in this book's GitHub repository at `https://github.com/PacktPublishing/Augmented-Reality-with-Unity-AR-Foundation`.

Exploring the AR Foundation example projects from Unity

A great way to learn about how to create AR projects with Unity AR Foundation is to explore the various example projects from Unity. These projects include example scenes, scripts, prefabs, and other assets. By cloning a project and opening an example scene, you can learn how to use AR Foundation, experiment with features, and see some best practices. In particular, consider these projects:

- *AR Foundation Samples*: `https://github.com/Unity-Technologies/arfoundation-samples`.
- *AR Foundation Demos*: `https://github.com/Unity-Technologies/arfoundation-demos`.
- *XR Interaction Toolkit Examples*: `https://github.com/Unity-Technologies/XR-Interaction-Toolkit-Examples/tree/master/AR`.
- For more advanced work, I'm also a fan of several individual contributors, including Dan Miller, a senior XR developer at Unity. See `https://github.com/DanMillerDev` for more information.

Please look through the **README** file for each of these projects (found on the GitHub project's home page) to gain an understanding of what the project does, any dependencies it has, and other useful information about the project.

Each of these repositories contains a full Unity project. That is, they are not simply Unity asset packages you can import into an existing project. Rather, you'll clone the entire repository and open it as its own project. This is typical for demo projects that may have other package dependencies and require preset settings to build and run properly.

The *AR Foundation Samples* project is my *go-to project* for learning various AR Foundation features. It contains many example scenes demoing individual features, often in place of detailed documentation elsewhere (see `https://github.com/Unity-Technologies/arfoundation-samples/tree/main/Assets/Scenes`).

Each scene is extremely simple (almost to a fault) as it has the atomic purpose of illustrating a single feature. For example, there are separate scenes for plane detection, plane occlusion, and feathered planes. Notably, the project also contains a main menu scene (`Assets/Scenes/ARFoundationMenu/Menu`) that launches the other scenes when you build them all into a single executable. I recommend starting with the scene named **SimpleAR**, which we'll review in a moment.

Another is the *AR Foundation Demos* project, which contains some more complex user scenarios and features not covered in the Samples project. For example, it demonstrates the Unity **Onboarding UX** assets, which we'll introduce you to in *Chapter 4, Creating an AR User Framework*. It also covers image tracking, mesh placement, language localization, and some useful shaders (for example, wireframe, shadows, and fog).

The *XR Interaction Toolkit Examples* repository contains two separate Unity projects: one for VR and another for AR. It is largely a placeholder (in my opinion) for things to come.

Information – XR Interaction Toolkit

The XR Interaction Toolkit from Unity is not covered in this book. It provides components and other assets for developing interactive scenes using hand controllers and device-supported hand gestures. At the time of writing, XR Interaction Toolkit is focused on **Virtual Reality** (**VR**) applications (evidenced by its Examples project, which contains seven scenes for VR and just one for AR, which only supports mobile AR) but I believe it is a key part of Unity's XR strategy and architecture for the future. If you are interested in XR Interaction Toolkit for VR, check out my other book, *Unity 2020 Virtual Reality Projects – Third Edition*, from Packt Publishing.

Let's get a copy of the AR Foundation Samples project and take a look at the **SimpleAR** scene.

Building and running the Samples project

In this section, you are going to build the *AR Foundation Samples* project and run it on your device. First, please clone the project from its GitHub repository and open it in Unity, as follows:

1. Clone a copy of the project from GitHub to your local machine. The project can be found at `https://github.com/Unity-Technologies/arfoundation-samples`. Please use whatever cloning method you prefer; for example, GitHub Desktop (`https://desktop.github.com/`) or the command line (`https://git-scm.com/download/`).

2. Add the project to **Unity Hub** by selecting **Projects | Add**, navigating to the cloned project's root folder, and pressing **Select Folder**.

3. Open the project in Unity. In the **Unity Hub** projects list, if you see a yellow warning icon, then the cloned project's Unity version is not currently installed on your system. Use the **Unity Version** selection to choose a newer version of the editor that you have, preferably of the same major release (for example, 20XX).

4. Open the project by selecting it from the Unity Hub projects list.

5. If your version of Unity is newer than the project from when it was last saved, you will see a prompt asking, "**Do you want to upgrade your project to a newer version of Unity?**." Press **Confirm**.

One of the scenes, **SimpleAR**, is a basic AR example scene. When run, the user will scan their room with their device's camera and the app will detect any horizontal planes that are rendered on the screen. When your user taps on one of these planes, a small red cube will be placed in the environment. You can walk around the room and the cube will remain where it was placed. If you tap again on another location, the cube will be moved there. Let's briefly review this **SimpleAR** scene's GameObjects:

1. Open the **SimpleAR** scene from the **Project** window by navigating to the `Scenes/SimpleAR/` folder and double-clicking the `SimpleAR` scene file.

2. In the **Hierarchy** window, you will find two GameObjects of particular interest: **AR Session** and **AR Session Origin**.

3. Select the **AR Session Origin** object and examine its components in the **Inspector** window. These include **AR Plane Manager**, **AR Point Cloud Manager**, **AR Raycast Manager**, and a **Place On Plane** script. We'll explain all of this later in this chapter.

Now, let's try to build and run the project:

1. Switch to your target platform if necessary. To do this, go to **File | Build Settings**, choose your device's platform from the **Platform** list (for example, Android or iOS), and click **Switch Platform**.

2. Most likely, the cloned project's settings have already been configured, but let's make sure. From the **Build Settings** window, click the **Player Settings** button to open that window and confirm the necessary settings mentioned in *Chapter 1, Setting Up for AR Development*. For example, Android ARCore does not support Vulcan graphics and needs **Nougat (API Level 24)** as a minimum requirement.

3. In the **Build Settings** window again, notice that the list of scenes in **Scenes in Build** starts with the **Menu** scene and contains all the demo scenes from this project (the first in the list will be the first scene to load when the app loads). You can leave these alone or just pick the one you want in the build.

4. Ensure your mobile device is plugged into a USB port on your computer.

5. Press the **Build And Run** button to build the project and install it on your device. It will prompt you for a file folder location; I like to create a folder in my project root, named `Builds/`. Give it a filename (if required) and press **Save**. It may take a while to complete this task.

If all goes well, the project will build, be installed on your device, and launch.

If you encounter errors while building the project, look at the **Console** window in the Unity Editor for messages (in the default layout, it's a tab behind the **Project** window). Read the messages carefully, generally starting from the top. If the fix is not obvious, do an internet search for the message's text, as you can be certain you're probably not the first person to have a similar question!

> **Tip – "Failed to generate ARCore reference image library" error**
>
> If you receive an error when attempting to build the project that says something like **Failed to generate ARCore reference image library**, please make sure there are no spaces in the pathname of your project folder! See `https://github.com/Unity-Technologies/arfoundation-samples/issues/119` for more information.

The main menu will be displayed, as shown in the following screen capture (left panel):

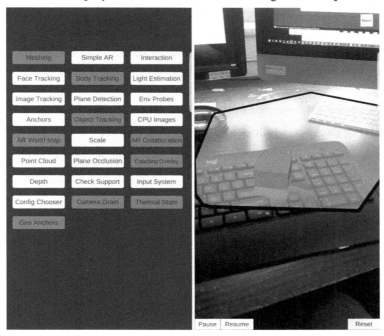

Figure 2.1 – Screenshot of my phone running the arfoundation-samples app and SimpleAR scene

A cool thing about AR Foundation (and this project) is that it can detect the capabilities of the device it is running on at runtime. This means that the buttons in the main menu will be disabled when AR Foundation detects that the features demoed in that scene are not supported on the device. (The device I'm using in the preceding screen capture is an Android phone, so some iOS-only feature scenes are disabled).

Click the **Simple AR** button to open that scene. You should see a camera video feed on your device's screen. Move your phone slowly in different directions and closer/away. As it scans the environment, feature points and planes will be detected and rendered on the screen. Tap one of the planes to place a cube on the scene, as shown in the right-hand panel of the preceding screen capture.

Some of the assets and scripts in the Samples project can be useful for building our own projects. I'll show you how to export them now.

Exporting the sample assets for reuse

Unity offers the ability to share assets between projects using .unitypackage files. Let's export the assets from the AR Foundation Samples project for reuse. One trick I like to do is move all the sample folders into a root folder first. With the *arfoundation-samples* project open in Unity, please perform the following steps:

1. In the **Project** window, create a new folder under Assets named ARF-samples by clicking the + icon (top left of the window) and selecting **Folder**.

2. Drag the following folders into the ARF-samples one: Materials, Meshes, Prefabs, Scenes, Scripts, Shaders, and Textures. That is, move all of them but leave the XR folder at the root.

3. Right-click on the ARF-samples folder and select **Export Package**.

4. The **Exporting Package** window will open. Click **Export**.

5. Choose a directory outside this project's root, name the file (for example, arf-samples), and click **Save**.

The `Assets/ARF-samples/` folder in the **Project** window is shown in the following screenshot:

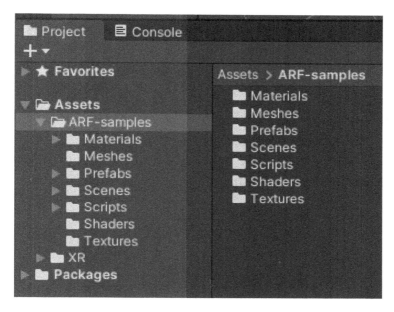

Figure 2.2 – The Samples assets folder being exported to a .unitypackage file

You can close the `arfoundation-samples` project now if you want. You now have an asset package you can use in other projects.

> **Tip – Starting a New Project by Copying the Samples Project**
>
> An alternative to starting a new Unity AR project from scratch is to duplicate the *arfoundation-samples* project as the starting point for new AR projects. To do that, from your Windows Explorer (or macOS Finder), duplicate the entire project folder and then add it to Unity Hub. This way, you get all the example assets and demo scenes in one place, and it's set up with reasonable default project settings. I often do this, especially for quick demos and small projects.

Next, we are going to import the Samples assets into your Unity project and build the given SimpleAR scene.

Building the SimpleAR scene in your own project

As you will see later in this chapter, the Samples project includes some assets we can use in your own projects, saving you time and effort, especially at the start. We will import `unitypackage`, which we just exported, and then build the given SimpleAR scene as another test to verify that you're set up to build and run AR applications.

Creating a new project

If you already have a Unity project set up for AR development, as detailed in *Chapter 1, Setting Up for AR Development*, you can open it in Unity and skip this section.
If not, perform the following steps, which have been streamlined for your convenience. If you require more details or explanations, please revisit *Chapter 1, Setting Up for AR Development*.

To create and set up a new Unity project with AR Foundation, Universal Render Pipeline, and the new Input System, here are the abbreviated steps:

1. *Create a new project* by opening **Unity Hub**, selecting **Projects | New**, choosing **Universal Render Pipeline**, specifying a **Project Name**, such as `MyARProject`, and clicking **Create**.

2. *Open your project* in the Unity Editor by selecting it from Unity Hub's **Projects** list.

3. *Set your target platform* by going to **File | Build Settings**, choosing **Android** or **iOS** from the **Platform** list, and clicking **Switch Platform**.

4. *Set up the Player Settings* according to *Chapter 1, Setting Up for AR Development*, and/or your device's documentation by going to the **Edit | Project Settings | Player** window. For example, Android ARCore does not support Vulcan graphics and needs **Nougat (API Level 24)** as a minimum requirement.

5. *Install an XR plugin* by going to **Edit | Project Settings | XR Plugins Manager | Install XR Plugin Management**. Then, check the checkbox for your device's **Plug-in Provider**.

6. *Install AR Foundation* by going to **Window | Package Manager**, choosing **Unity Registry** from the filter list at the top left, searching for `ar` using the search input field, selecting the **AR Foundation** package, and clicking **Install**.

7. *Install the Input System package* by going to **Window | Package Manager**, choosing **Unity Registry** from the filter list at the top left, searching for input using the search input field, selecting the **Input System** package, and clicking **Install**.

   When prompted to enable the input backend, you can say **Yes**, but we'll actually change this setting to **Both** in the next topic when we import the Sample assets into the project.

8. *Add the AR Background Renderer* to the URP Forward renderer by locating the **ForwardRenderer** data asset, usually in the Assets/Settings/ folder. In its **Inspector** window, click **Add Renderer Feature** and select **AR Background Renderer Feature**.

You might want to bookmark these steps for future reference. Next, we'll import the Sample assets we exported from the AR Foundation Samples project.

Importing the Sample assets into your own project

Now that you have a Unity project set up for AR development, you can import the sample assets into your project. With your project open in Unity, perform the following steps:

1. Import the package from the main menu by selecting **Assets | Import Package | Custom Package**.

2. Locate the arf-samples.unitypackage file on your system and click **Open**.

3. The **Import Unity Package** window will open. Click **Import**.

4. If you created your project using the *Universal Render Pipeline* (or HDRP), rather than using the built-in render pipeline like we did, you need to convert the imported materials. Select **Edit | Render Pipeline | URP | Upgrade Project Materials to URP Materials**. Then, when prompted, click **Proceed**.

5. Then, go to **Player Settings** using **Edit | Project Settings | Player**, select **Configuration | Active Input Handling**, and choose **Both**. Then, when prompted, click **Apply**.

6. We will use the new Input System for projects in this book. However, some demo scenes in the Samples project use the old Input Manager. If you choose **Input System Package (New)** for **Active Input Handling**, then those demo scenes may not run.

Hopefully, all the assets will import without any issues. However, there may be some errors while compiling the Samples scripts. This could happen if the Samples project is using a newer version of AR Foundation than your project and it is referencing API functions for features your project does not have installed. The simplest solution is to upgrade the version of AR Foundation to the same or later version as the Samples project. To do so, perform the following steps:

1. To see error messages, open the **Console** window using its tab or selecting **Window | General | Console**.

2. Suppose that, in my project, I have additional errors because I have installed *AR Foundation 4.0.12* but the Samples project uses *version 4.1.3* features, which are not available in my version. Here, I'll go to **Window | Package Manager**, select the **AR Foundation** package, click **See Other Versions**, select the 4.1.3 version, and then click the **Update to 4.1.3** button.

3. The project also might be using preview versions of packages. Enable preview packages by selecting **Edit | Project Settings | Package Manager | Enable preview packages**.

4. Ensure the ARCore XR plugin and/or AR Kit XR plugin version matches the version of the AR Foundation package the project is using.

5. Another message you might see is that some Samples scripts require that you enable "unsafe" code in the project. Go to **Project Settings | Player | Script Compilation | Allow 'unsafe' code** and check the checkbox.

    This is not as threatening as it may sound. "Unsafe" code usually means that something you installed is calling C++ code from the project that is *potentially* unsafe from the compiler's point of view. Enabling unsafe code in Unity is usually not a problem unless, for example, you are publishing WebGL to a WebPlayer, which we are not.

Finally, you can verify your setup by building and running the SimpleAR scene, this time from your own project. Perform the following steps:

1. Open the **SimpleAR** scene from the **Project** window by navigating to the `ARF-samples/Scenes/SimpleAR/` folder and double-clicking the **SimpleAR** scene file.

2. Open the **Build Settings** window by going to **File | Build Settings**.

3. For the **Scenes in Build** list, click the **Add Open Scenes** button and uncheck all the scenes in the list other than the SimpleAR one.

4. Ensure your device is connected via USB.

5. Press the **Build And Run** button to build the project and install it on your device. It will prompt you for a location; I like to create a folder in my project root named `Builds/`. Give it a filename (if required) and press **Save**. It may take a while to complete this task.

The app should successfully build and run on your device. If you encounter any errors, please review each of the steps detailed in this chapter and *Chapter 1, Setting Up for AR Development*.

When the app launches, as described earlier, you should see a camera video feed on your screen. Move your phone slowly in different directions and closer/away. As it scans the environment, feature points and planes will be detected and rendered on the screen. Tap one of these planes to place a cube on the scene.

Your project is now ready for AR development!

Starting a new, basic AR scene

In this section, we'll create a scene very similar to `SimpleAR` (actually, more like the Samples scene named `InputSystem_PlaceOnPlane`) but we will start with a new empty scene. We'll add AR Session and AR Session Origin objects provided by AR Foundation to the scene hierarchy, and then add trackable feature managers for planes and point clouds. In the subsequent sections of this chapter, we'll set up an Input System action controller, write a C# script to handle any user interaction, and create a prefab 3D graphic to place in the scene.

So, start the new scene by performing the following steps:

1. Create a new scene by going to **File | New Scene**.

2. If prompted, choose the **Basic (Built-in)** template. Then, click **Create**.

    Unity allows you to use a Scene template when creating a new scene. The one named **Basic (Built-in)** is comparable to the default new scene in previous versions of Unity.

3. Delete **Main Camera** from the **Hierarchy** window by using *right-click* | **Delete** (or the *Del* key on your keyboard).

4. Add an AR Session by selecting **GameObject** from the main menu, then **XR | AR Session**.

5. Add an AR Session Origin by selecting **GameObject** from the main menu, then **XR | AR Session Origin**.

6. Unfold **AR Session Origin** and select its child; that is, **AR Camera**. In the **Inspector** window, use the **Tag** selector at the top left to set it as our **MainCamera**. (This is not required but it is a good practice to have one camera in the scene tagged as **MainCamera**.)

7. Save the scene using **File | Save As**, navigate to the `Assets/Scenes/` folder, name it `BasicARScene`, and click **Save**.

Your scene Hierarchy should now look as follows:

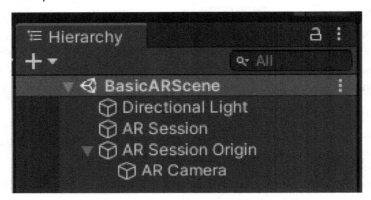

Figure 2.3 – Starting a scene Hierarchy

We can now take a closer look at the objects we just added, beginning with the AR Session object.

Using AR Session

The **AR Session** object is responsible for enabling and disabling augmented reality features on the target platform. When you select the **AR Session** object in your scene **Hierarchy**, you can see its components in the **Inspector** window, as shown in the following screenshot:

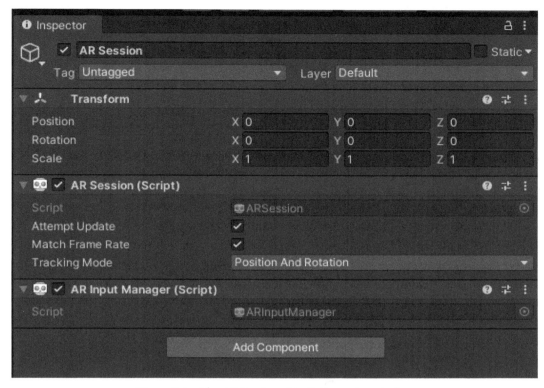

Figure 2.4 – The AR Session object's Inspector window

Each AR scene must include one (and only one) AR Session. It provides several options. Generally, you can leave these as their default values.

The **Attempt Update** option instructs the AR Session to try and install the underlying AR support software on the device if it is missing. This is not required for all devices. iOS, for example, does not require any additional updates if the device supports AR. On the other hand, to run AR apps on Android, the device must have the ARCore services installed. Most AR apps will do this for you if they are missing, and that is what the **Attempt Update** feature of **AR Session** does. If necessary, when your app launches and support is missing or needs an update, AR Session will attempt to install *Google Play Services for AR* (see `https://play.google.com/store/apps/details?id=com.google. ar.core`). If the required software is not installed, then AR will not be available on the device. You could choose to disable automatic updates and implement them yourself to customize the user onboarding experience.

> **Note**
> The **Match Frame Rate** option in the **Inspector** window is obsolete. Ordinarily, you would want the frame updates of your apps to match the frame rate of the physical device, and generally, there is no need to tinker with this. If you need to tune it, you should control it via scripting (see `https://docs.unity3d.com/ScriptReference/Application-targetFrameRate.html`).

Regarding **Tracking Mode**, you will generally leave it set to **Position and Rotation**, as this specifies that your VR device is tracking in the physical world 3D space using both its XYZ position and its rotation around each axis. This is referred to as *6DOF*, for six-degrees-of-freedom tracking, and is probably the behavior that you expect. But for face tracking, for example, we should set it to **Rotation Only**, as you'll see in *Chapter 9, Selfies: Making Funny Faces*.

The **AR Session** GameObject also has an **AR Input Manager** component that manages our **XR Input Subsystem** for tracking the device's pose in a physical 3D space. It reads input from the AR Camera's **AR Pose Driver** (discussed shortly). There are no options for the component, but this is required for device tracking.

We also added an AR Session Origin GameObject to the Hierarchy. Let's look at that next.

Using AR Session Origin

The **AR Session Origin** will be the root object of all trackable objects. Having a root origin keeps the Camera and any trackable objects in the same space and their positions relative to each other. This *session* (or *device) space* includes the AR Camera and any *trackable* features that have been detected in the real-world environment by the AR software. Otherwise, detected features, such as planes, won't appear in the correct place relative to the Camera.

Tip – Scaling Virtual Scenes in AR

If you plan to scale your AR scene, place your game objects as children of AR Session Origin and then scale the parent AR Session Origin transform, rather than the child objects themselves. For example, consider a world-scale city map or game court resized to fit on a tabletop. Don't scale the individual objects in the scene; instead, scale everything by resizing the root session origin object. This will ensure the other Unity systems, especially physics and particles, retain their scale relative to the camera space. Otherwise, things such as gravity, calculated as meters per second, and particle rendering could mess up.

When you select the **AR Session Origin** object in your scene **Hierarchy**, you can see its components in the **Inspector** window, as shown in the following screenshot:

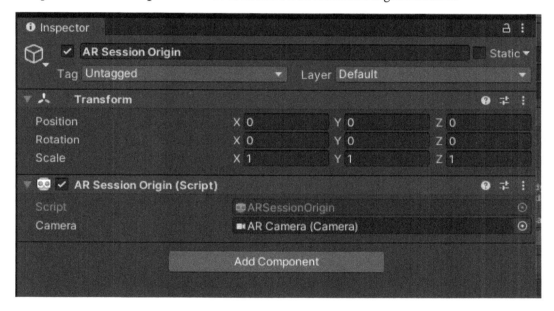

Figure 2.5 – The AR Session object's Inspector window

At the time of writing, the default AR Session Origin object simply has an **AR Session Origin** component. We'll want to build out its behavior by adding more components in a moment.

The Session Origin's **Camera** property references its own child **AR Camera** GameObject, which we'll look at next.

Using the AR Camera

The **AR Camera** object is a child of AR Session Origin. Its **Inspector** window is shown in the following screenshot:

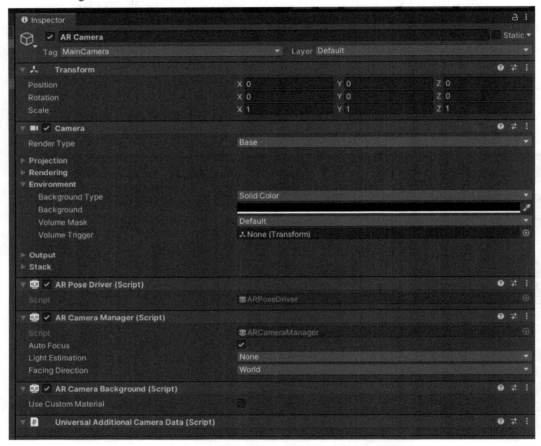

Figure 2.6 – The AR Camera object's Inspector window

During setup, we tagged the AR Camera as our **MainCamera**. This is not required but it is a good practice to have one camera in the scene tagged as MainCamera, for example, for any code that may use Camera.main, which is a shortcut for the find by tag name.

As its name implies, the AR Camera object includes a **Camera** component, required in all Unity scenes, which determines what objects to render on your screen. The AR one has mostly default values. The **Near** and **Far Clipping** planes have been adjusted for typical AR applications to (0.1, 20) meters. In AR apps, it's not unusual to place the device within inches of a virtual object, so we wouldn't want it to be clipped. Conversely, in an AR app, if you walk more than 20 meters away from an object that you've placed in the scene, you probably don't need it to be rendered at all.

Importantly, rather than using a Skybox, as you'd expect in non-AR scenes, the camera's **Background** is set to a **Solid** black color. This means the background will be rendered using the camera's video feed. This is controlled using the **AR Camera Background** component of the AR Camera. In an advanced application, you can even customize how the video feed is rendered, using a custom video *material* (this topic is outside the scope of this book). Similarly, on a wearable AR device, a black camera background is required, but with no video feed, to mix your virtual 3D graphics atop the visual see-through view.

The video feed source is controlled using the AR Camera Manager component. You can see, for example, that **Facing Direction** can be changed from **World** to **User** for a selfie face tracking app (see *Chapter 9, Selfies: Making Funny Faces*).

The **Light Estimation** options are used when you want to emulate real-world lighting when rendering your virtual objects. We'll make use of this feature later in this chapter.

You also have the option to disable **Auto Focus** if you find that the camera feature is inappropriate for your AR application.

Tip – When to Disable Camera Auto Focus for AR

Ordinarily, I disable Auto Focus for AR applications. When the software uses the video feed to help detect planes and other features in the environment, it needs a clear, consistent, and detailed video feed, not one that may be continually changing for Auto Focus. That would make it difficult to process AR-related algorithms accurately to decode their tracking. On the other hand, a selfie face tracking app may be fine with Auto Focus enabled and could improve the user experience when the area behind the user loses focus due to depth of field.

The **AR Pose Driver** component is responsible for updating the AR Camera's transform as it tracks the device in the real world. (There are similar components for VR headsets and hand controllers, for instance.) This component relies on the XR plugin and the Input XR Subsystem to supply the positional tracking data (see `https://docs.unity3d.com/Manual/XRPluginArchitecture.html`).

Our next step is to add Plane and Point Cloud visualizers to the scene.

Adding Plane and Point Cloud managers

When your application runs, you'll ask the user to scan the room for the AR software to detect features in the environment, such as depth points and flat planes. Usually, you'll want to show these to the user as they're detected. We do this by adding the corresponding feature managers to the AR Session Origin game object. For example, to visualize planes, you'll add an **AR Plane Manager** to the AR Session Origin object, while to visualize point clouds, you'll add an **AR Point Cloud Manager**.

AR Foundation supports detecting and tracking the following features:

- *Anchor*: A fixed pose (consisting of location and rotation) in the physical environment (controlled by the AR Anchor Manager component). This is also known as a Reference Point.

- *Reflection Probe*: Environment reflection probes for rendering shiny surface materials (controlled by the AR Environment Probe Manager component).

- *Face*: A human face detected by the AR device (controlled by the AR Face Manager component).

- *Human Body*: A trackable human body and the body's skeleton (controlled by the AR Human Body Manager component).

- *Image*: A 2D image that has been detected and tracked in the environment's AR Tracked Image Manager component.

- *Participant*: Another user (device) in a collaborative session.

- *Plane*: A flat plane, usually horizontally or vertically inferred from the point cloud (controlled by the AR Plane Manager component).

- *Point Cloud*: A set of depth points detected by the AR device (controlled by the AR Point Cloud Manager component).

- *Object*: A 3D object detected and tracked in the environment (controlled by the AR Tracked Object Manager component).

Not all of these are supported on every platform. See the documentation for your current version of AR Foundation (for example, visit `https://docs.unity3d.com/Packages/com.unity.xr.arfoundation@4.1/manual/index.html#platform-support` and select your version at the top left). We will be using many of these in various projects throughout this book. Here, we will use the Plane and Point Cloud trackables. Please perform the following steps to add them:

1. Select the **AR Session Origin** object from the **Hierarchy** window.

2. Add a Point Cloud Manager by selecting **Add Component**, searching for `ar` in the search input field, then clicking **AR Point Cloud Manager**.

3. Add a Plane Manager by selecting **Add Component**, searching for `ar` in the search input field, and clicking **AR Plane Manager**.

4. On the AR Plane Manager, change **Detection Mode** to only horizontal planes by selecting **Nothing** (to clear the list), then selecting **Horizontal**.

You'll notice that the Point Cloud Manager has an empty slot for the Point Cloud Prefab visualizer and that the Plane Manager has an empty slot for the Plane Prefab visualizer. We'll use prefabs from the Samples project, as follows:

1. In the **Inspector** window, go to **AR Point Cloud Manager | Point Cloud Prefab** and press the *doughnut* icon on the right-hand side of the field to open the **Select GameObject** dialog box.

2. Click the **Assets** tab and double-click the **AR Point Cloud Visualizer** prefab.

   There are alternative point cloud visualizer prefabs you might like to try out also, such as **AR Point Cloud Debug Visualizer** and **AllPointCloudPointsPrefab**.

3. Likewise, for **AR Plane Manager | Plane Prefab**, press the *doughnut* icon on the right-hand side of the field to open the **Select GameObject** dialog box.

4. Click the **Assets** tab and double-click **AR Feathered Plane**.

   There are alternative plane visualizer prefabs to try out also, such as **AR Plane Debug Visualizer**, **AR Feathered Plane Fade**, and **CheckeredPlane**.

5. Save the scene by going to **File | Save**.

We're using the visualizer prefabs we got from the Samples project. Later in this chapter, we'll talk more about prefabs, take a closer look at the visualizer ones, and learn how to edit them to make our own custom visualizers. First, we'll add the AR Raycast Manager to the scene.

Adding AR Raycast Manager

There's another component I know we're going to need soon, known as **AR Raycast Manager**. This will be used by our scripts to determine if a user's screen touch corresponds to a 3D trackable feature detected by the AR software. We're going to use it in our script to place an object on a plane. Perform the following steps to add it to the scene:

1. Select the **AR Session Origin** object from the **Hierarchy** window.

2. Select **Add Component |** search for `ar` in the search input field, and click **AR Raycast Manager**.

The **AR Session Origin** GameObject with the manager components we added now looks like this in the **Inspector** window:

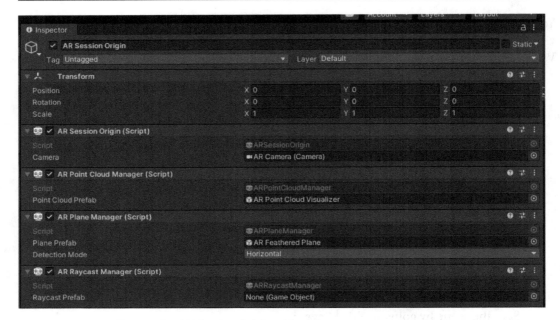

Figure 2.7 – AR Session Origin with various manager components

One more thing that's handy to include is light estimation, which helps with rendering your virtual objects more realistically.

Adding Light Estimation

By adding a Light Estimation component to your Directional Light source, the AR camera can use this information when rendering your scene to try and match the scene's lighting more closely to the real-world environment.

To add light estimation, perform the following steps:

1. In the **Hierarchy** window, select the **Directional Light** object.

2. In the **Inspector**, click **Add Component**, search for light estimation, and add the **Basic Light Estimation** component.

3. In the **Hierarchy** window, find **AR Camera** (child of AR Session Origin), drag it into the **Inspector** window, and drop it onto the **Light Estimation | Camera Manager** slot.

4. In the **Hierarchy** window, select **AR Camera**, then set **AR Camera Manager | Light Estimation** to **Everything**. Note that not all platforms support all light estimation capabilities, but using the **Everything** flags will have them use all of the ones that are available at runtime.

5. Save your work by going to **File | Save**.

Good! I think we should try to build and run what we have done so far and make sure it's working.

Building and running the scene

Currently, the scene initializes an AR Session, enables the AR camera to scan the environment, detects points and horizontal planes, and renders these on the screen using visualizers. Let's build the scene and make sure it runs:

1. Open the **Build Settings** window by going to **File | Build Settings**.
2. For the **Scenes in Build** list, click the **Add Open Scenes** button and uncheck all the scenes in the list other than this current scene (mine is named **BasicARScene**).
3. Ensure your device is connected to your computer via USB.
4. Press the **Build And Run** button to build the project and install it on your device. It will prompt you for a location; I like to create a folder in my project root named `Builds/`. Give it a filename (if required) and press **Save**. It may take a while to complete this task.

The app should successfully build and run on your device. If you encounter any errors, please read the error messages carefully in the **Console** window. Then, review each of the setup steps detailed in this chapter and *Chapter 1, Setting Up for AR Development*.

When the app launches, you should see a video feed on your screen. Move the device slowly in different directions and closer/away. As it scans the environment, feature points and planes will be detected and rendered on the screen using the visualizers you chose.

Next, let's add the ability to tap on one of the planes to instantiate a 3D object there.

Placing an object on a plane

We will now add the ability for the user to tap on a plane and place a 3D virtual object in the scene. There are several parts to implementing this:

- Setting up a Place Object input action when the user taps the screen.
- Writing a PlaceObjectOnPlane script that responds to the input action and places an object on the plane.
- Determining which plane and where to place the object using AR Raycast Manager.
- Importing a 3D model and making it a prefab for placing in this scene.

Let's begin by creating an input action for a screen tap.

Setting up a PlaceObject input action

We are going to use the Unity Input System package for user input. If the Input System is new to you, the steps in this section may seem complicated, but only because of its great versatility.

The Input System lets you define **Actions** that separate the logical meaning of the input from the physical means of the input. Using named actions is more meaningful to the application and programmers.

> **Note – Input System Tutorial**
>
> For a more complete tutorial on using the Input System package, see
> `https://learn.unity.com/project/using-the-input-system-in-unity`.

Here, we will define a **PlaceObject** action that is bound to screen tap input data. We'll set this up now, and then use this input action in the next section to find the AR plane that was tapped and place a virtual object there.

Before we begin, I will assume you have already imported the **Input System** package via **Package Manager** and set **Active Input Handing** to **Input System Package** (or **Both**) in **Player Settings**. Now, follow these steps:

1. In the **Project** window, create a new folder named `Inputs` using *right-click* | **Create | Folder** (or use the + button at the top left of the window). I put mine under my `_App/` folder.

2. Create an input action controller asset by *right-clicking* inside the `Inputs` folder, then selecting **Create | Input Actions** (or using the + button at the top left of the window). Rename it `AR Input Actions`.

3. Click **Edit Asset** to open its editor window.

4. In the leftmost **Action Maps** panel, click the + button and name the new map `ARTouchActions`.

5. In the middle **Actions** panel, rename the default action to `PlaceObject` using *right-click* | **Rename**.

6. In the right-hand side **Properties** panel, set **Action Type** to **Value**.

7. Set its **Control Type** to **Vector 2**.

8. In the middle **Actions** panel, click the child **<No Binding>** item to add a binding.

9. In the right-hand side **Properties** panel, under **Binding**, using the **Path** select list, choose **TouchScreen | Primary Touch | Position**.

10. At the top of the window, click **Save Asset** (unless the **Auto-Save** checkbox is checked).

With that, we've created a data asset named **AR Input Actions** that contains an action map named **ARTouchActions**, which has one action, **PlaceObject**, that detects a screen touch. It returns the touch position as a 2D vector (Vector2) with the X, Y values in pixel coordinates. The input action asset is shown in the following screenshot:

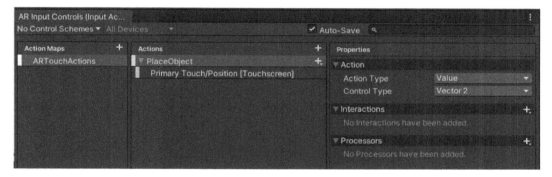

Figure 2.8 – Our AR Input Actions set up for screen taps

Now, we can add the input actions to the scene. This can be done via a Player Input component. For our AR scene, we'll add a Player Input component to the AR Session Origin, as follows:

1. In the **Hierarchy** window, select the **AR Session Origin** object.

2. In its **Inspector** window, click **Add Component | Input | Player Input**.

3. From the **Project** window, drag the **AR Input Actions** asset from your Inputs/ folder into the **Player Input | Actions** slot in the **Inspector** window.

4. Leave **Behavior** set to **Send Messages**.

> **Information – Input System Behavior Types**
>
> Unity and C# provide different ways for objects to signal other objects. The Player Input component lets you choose how you want input actions to be communicated, via its **Behavior** setting. The options are as follows:
>
> *Send Messages*: Will send action messages to any components on the *same* GameObject (`https://docs.unity3d.com/ScriptReference/GameObject.SendMessage.html`). As we'll see, your message handler must be named with the "On" prefix (for example, `OnPlaceObject`) and receives an `InputValue` argument (`https://docs.unity3d.com/Packages/com.unity.inputsystem@1.1/api/UnityEngine.InputSystem.InputValue.html`).
>
> *Broadcast Messages*: Like Send Messages, Broadcast Messages will send messages to components on this GameObject *and all its children* (`https://docs.unity3d.com/ScriptReference/Component.BroadcastMessage.html`).
>
> *Invoke Unity Events*: You can set event callback functions using the Inspector or in scripts (`https://docs.unity3d.com/Manual/UnityEvents.html`). The callback function receives an `InputAction.CallbackContext` argument (`https://docs.unity3d.com/Packages/com.unity.inputsystem@1.1/api/UnityEngine.InputSystem.InputAction.CallbackContext.html`).
>
> *Invoke C# Events*: You can set event listeners in scripts (`https://docs.microsoft.com/en-us/dotnet/csharp/programming-guide/events/`).
>
> To learn more about the Player Input component, see `https://docs.unity3d.com/Packages/com.unity.inputsystem@1.0/api/UnityEngine.InputSystem.PlayerInput.html`.

I've decided to use **Send Messages** here, so we'll need to write a script with an `OnPlaceObject` function, which we'll do next. But first, I'll provide a quick introduction to Unity C# programming.

Introducing Unity C# programming and the MonoBehaviour class

Writing C# scripts is an essential skill for every Unity developer. You don't need to be an expert programmer, but you cannot avoid writing some code to make your projects work. If you are new to coding, you can simply follow the instructions provided here, and over time, you'll get more comfortable and proficient. I also encourage you to go through some of the great beginner tutorials provided by Unity (`https://learn.unity.com/`) and others, including the following:

- **Coding in C# in Unity for Beginners**: `https://unity3d.com/learning-c-sharp-in-unity-for-beginners`

- **Working with Scripts**: `https://learn.unity.com/tutorial/working-with-scripts`

- **Beginner Scripting**: `https://learn.unity.com/project/beginner-gameplay-scripting`

Given that, I will offer some brief explanations as we work through this section. But I'll assume that you have at least a basic understanding of C# language syntax, common programming vocabulary (for example, *class*, *variable*, and *function*), using an editor such as Visual Studio, and how to read error messages that may appear in your **Console** window due to typos or other common coding mistakes.

We're going to create a new script named `PlaceObjectOnPlane`. Then, we can attach this script as a component to a GameObject in the scene. It will then appear in the object's **Inspector** window. Let's begin by performing the following steps:

1. In the **Project** window, locate your `Scripts/` folder (mine is `Assets/_App/Scripts/`), *right-click* it, and select **Create | C# Script**.

2. Name the file `PlaceObjectOnPlane` (no spaces nor other special characters are allowed in the name, and it should start with a capital letter).

   This creates a new C# script with the `.cs` file extension (although you don't see the extension in the **Project** window).

3. Double-click the **PlaceObjectOnPlane** file to open it in your code editor. By default, my system uses Microsoft Visual Studio.

As you can see in the following initial script content of the template, the `PlaceObjectOnPlane.cs` file declares a C# class, `PlaceObjectsOnPlane`, that has the same name as the `.cs` file (the names *must* match; otherwise, it will cause compile errors in Unity):

```
using System.Collections;
using System.Collections.Generic;
using UnityEngine;


public class PlaceObjectOnPlane : MonoBehaviour
{
    // Start is called before the first frame update
    void Start()
```

```
    {
    }

    // Update is called once per frame
    void Update()
    {
    }
}
```

The first three lines in this script have a using directive, which declares an SDK library, or namespace, that will be used in the script. When a script references external symbols, the compiler needs to know where to find them. In this case, we're saying that we'll potentially be using standard .NET system libraries for managing sets of objects (*collections*). And here, we are using the UnityEngine API.

One of the symbols defined by *UnityEngine* is the **MonoBehaviour** class. You can see that our PlaceObjectsOnPlane class is declared as a subclass of MonoBehaviour. (Beware its British spelling, "iour"). Scripts attached to a GameObject in your scene must be a subclass of MonoBehaviour, which provides a litany of features and services related to the GameObject where it is attached.

For one, MonoBehaviour provides hooks into the GameObject life cycle and the Unity **game loop**. When a GameObject is created at runtime, for example, its Start() function will automatically be called. This is a good place to add some initialization code.

The Unity game engine's main purpose is to render the current scene view every frame, perhaps 60 times per second or more. Each time the frame is updated, your Update() function will automatically be called. This is where you put any runtime code that needs to be run every frame. Try to keep the amount of work that's done in Update() to a minimum; otherwise, your app may feel slow and sluggish.

You can learn more about the MonoBehaviour class here: https://docs.unity3d. com/ScriptReference/MonoBehaviour.html. To get a complete picture of the GameObject and MonoBehaviour scripts' life cycles, take a look at this flowchart here: https://docs.unity3d.com/Manual/ExecutionOrder.html.

We can now write our script. Since this is the first script in this book, I'll present it slowly.

Writing the PlaceObjectOnPlane script

The purpose of the `PlaceObjectOnPlane` script is to place a virtual object on the AR plane when and where the user taps. We'll outline the logic first (in C#, any text after `//` on the same line is a comment):

```csharp
using System.Collections;
using System.Collections.Generic;
using UnityEngine;
using UnityEngine.InputSystem;

public class PlaceObjectOnPlane : MonoBehaviour
{
    void OnPlaceObject(InputValue value)
    {
        // get the screen touch position
        // raycast from the touch position into the 3D scene
        //    looking for a plane
        // if the raycast hit a plane then
        //       get the hit point (pose) on the plane
        //       if this is the first time placing an object,
        //           instantiate the prefab at the hit position
        //           and rotation
        //       else
        //           change the position of the previously
        //           instantiated object

    }
}
```

As it turns out, in this script, there is no need for an `Update` function as it is only used for frame updates, which this script can ignore.

This script implements `OnPlaceObject`, which is called when the user taps the screen. As we mentioned previously, the Player Input component we added to the AR Session Origin uses the **Send Messages** behavior and thus expects our script to implement `OnPlacedObject` for the **PlacedObject** action. It receives an `InputValue`. Notice that I also added a line using `UnityEngine.InputSystem;`, which defines the `InputValue` class.

First, we need to get the screen touch position from the input value we passed in. Add the following code, which declares and assigns it to the `touchPosition` local variable:

```
// get the screen touch position
Vector2 touchPosition = value.Get<Vector2>();
```

The next step is to figure out if the screen touch corresponds to a plane that was detected in the AR scene. AR Foundation provides a solution by using the AR Raycast Manager component that we added to the AR Session Origin GameObject earlier. We'll use it in our script now. Add these lines to the top of your script:

```
using UnityEngine.XR.ARFoundation;
using UnityEngine.XR.ARSubsystems;
```

Then, inside the `OnPlaceObject` function, add the following code:

```
// raycast from the touch position into the 3D scene
//   looking for a plane
// if the raycast hit a plane then
ARRaycastManager raycaster =
    GetComponent<ARRaycastManager>();
List<ARRaycastHit> hits = new List<ARRaycastHit>();


if (raycaster.Raycast(touchPosition, hits,
    TrackableType.PlaneWithinPolygon))
{
    //
}
```

Firstly, we get a reference to the **ARRaycastManager** component, assigning it to `raycaster`. We declare and initialize a list of `ARRaycastHit`, which will be populated when the raycast finds something. Then, we call `raycaster.Raycast()`, passing in the screen's `touchPosition`, and a reference to the `hits` list. If it finds a plane, it'll return `true` and populate the `hits` list with details. The third argument instructs `raycaster.Raycast` on what kinds of trackables can be hit. In this case, `PlaneWithinPolygon` filters for 2D convex-shaped planes.

> **Information – For More Information on AR Raycasting**
>
> For more information on using ARRaycastManager, see `https://docs.unity3d.com/Packages/com.unity.xr.arfoundation@4.1/manual/raycast-manager.html`.
>
> For a list of trackable types you can pass in, see `https://docs.unity3d.com/Packages/com.unity.xr.arsubsystems@4.1/api/UnityEngine.XR.ARSubsystems.TrackableType.html`.

The code inside the `if` statement will only be executed if `raycaster.Raycast` returns `true`; that is, if the user had tapped a location on the screen that casts to a trackable plane in the scene. In that case, we must create a 3D GameObject there. In Unity, creating a new GameObject is referred to as **instantiating** the object. You can read more about it here: `https://docs.unity3d.com/Manual/InstantiatingPrefabs.html`.

First, let's declare a variable, `placedPrefab`, to hold a reference to the prefab that we want to instantiate on the selected plane. Using the `[SerializedField]` directive permits the property to be visible and settable in the Unity Inspector. We'll also declare a `private` variable, `spawnedObject`, that holds a reference to the instantiated object. Add the following code to the top of the class:

```
public class PlaceObjectOnPlane : MonoBehaviour
{
    [SerializeField] GameObject placedPrefab;
    GameObject spawnedObject;
```

Now, inside the `if` statement, we will instantiate a new object if this is the first time the user has tapped the screen, and then assign it to `spawnedObject`. If the object had already been spawned and the user taps the screen again, we'll move the object to the new location instead. Add the following highlighted code:

```
public void OnPlaceObject(InputValue value)
{
    // get the screen touch position
    Vector2 touchPosition = value.Get<Vector2>();

    // raycast from the touch position into the 3D scene
        looking for a plane
    // if the raycast hit a plane then
```

```
ARRaycastManager raycaster =
    GetComponent<ARRaycastManager>();
List<ARRaycastHit> hits = new List<ARRaycastHit>();

if (raycaster.Raycast(touchPosition, hits,
    TrackableType.PlaneWithinPolygon))
{
    // get the hit point (pose) on the plane
    Pose hitPose = hits[0].pose;

    // if this is the first time placing an object,
    if (spawnedObject == null)
    {
        // instantiate the prefab at the hit position
            and rotation
        spawnedObject = Instantiate(placedPrefab,
            hitPose.position, hitPose.rotation);
    }
    else
    {
        // change the position of the previously
            instantiated object
        spawnedObject.transform.SetPositionAndRotation(
            hitPose.position, hitPose.rotation);
    }
}
```

Raycast populates a list of hit points, as there could be multiple trackable planes in line where the user has tapped the screen. They're sorted closest to furthest, so in our case, we're only interested in the first one, hits[0]. From there, we get the point's Pose, a simple structure that includes 3D position and rotation values. These, in turn, are used when placing the object.

After that, save the script file.

Now, back in Unity, we'll attach our script as a component to **AR Session Origin** by performing the following steps:

1. First, check the **Console** window (using the **Console** tab or **Window | General | Console**) and ensure there are no compile errors from the script. If there are, go back to your code editor and fix them.

2. In the **Hierarchy** window, select the **AR Session Origin** object.

3. In the **Project** window, drag the **PlaceObjectOnPlane** script into the **Inspector** window so that when you drop it, it is added as a new component.

    You'll notice that there is a Placed Prefab property in the component's **Inspector** window. This is the `placedPrefab` variable we declared in the script. Let's populate it with the red cube prefab provided by the Samples assets.

4. In the **Project** window, navigate to the `ARF-samples/Prefabs/` folder.

5. Drag the **AR Placed Cube** prefab into the **Inspector** window, on the **Place Object On Plane | Placed Prefab** slot.

6. Save the scene by going to **File | Save**.

Our script, as a component on the AR Session Origin GameObject, should now look as follows:

Figure 2.9 – PlaceObjectOnPlane as a component with its Placed Prefab slot populated

Let's try it! We're now ready to build and run the scene.

Building and running the scene

If you've built the scene before, in the previous section, you can go to **File | Build And Run** to start the process. Otherwise, perform the following steps to build and run the scene:

1. Open the **Build Settings** window by going to **File | Build Settings**.

2. For the **Scenes in Build** list, click the **Add Open Scenes** button and uncheck all the scenes in the list other than this one (mine is named **BasicARScene**).

3. Ensure your device is connected via USB.

4. Press the **Build And Run** button to build the project and install it on your device. It will prompt you for a location; I like to create a folder in my project root named `Builds/`. Give it a filename (if required) and press **Save**. It may take a while to complete this task.

The app should successfully build and run on your device. As usual, if you encounter any errors, please read the error messages carefully in the **Console** window. When the app launches, you should see a video feed on your screen. Move your device slowly in different directions and closer/away. As it scans the environment, feature points and planes will be detected and rendered on the screen. If you tap the screen on a tracked plane, the red cube should be placed at that location.

Refactoring your script

Refactoring is reworking a script to make the code cleaner, more readable, more organized, more efficient, or otherwise improved without changing its behavior or adding new features. We can now refactor our little script to make the following improvements:

- Move initialization code that only needs to be done once out of `Update()` into `Start()` (for example, initialize the `raycaster` variable).

- Avoid allocating new memory in `Update()` to avoid memory fragmentation and garbage collection (for example, initialize the `hits` list as a class variable).

The modified script is shown in the following code block. The changed code is highlighted, beginning with the top part, which contains the new class variables and the `Start()` function:

```
public class PlaceObjectOnPlane : MonoBehaviour
{

    [SerializeField] GameObject placedPrefab;


    GameObject spawnedObject;
    ARRaycastManager raycaster;
    List<ARRaycastHit> hits = new List<ARRaycastHit>();


    void Start()
    {
        raycaster = GetComponent<ARRaycastManager>();
    }
```

Now, add the `OnPlacedObject` function, as follows:

```
public void OnPlaceObject(InputValue value)
{
    // get the screen touch position
    Vector2 touchPosition = value.Get<Vector2>();

    // raycast from the touch position into the 3D scene
        looking for a plane
    // if the raycast hit a plane then
    // REMOVE NEXT TWO LINES
    // ARRaycastManager raycaster =
        GetComponent<ARRaycastManager>();
    //List<ARRaycastHit> hits = new List<ARRaycastHit>();
```

5.   if (raycaster.Raycast(touchPosition, hits,
        TrackableType.PlaneWithinPolygon))

```
        {
```

Please save the script, then build and run it one more time to verify it still works.

Information – Public versus Private and Object Encapsulation

One of the driving principles of object-oriented programming is **encapsulation**, where an object keeps its internal variables and functions private, and only exposes properties (public variables) and methods (public functions) to other objects when they're intended to be accessible. C# provides the `private` and `public` declarations for this purpose. And in C#, any symbol not declared public is assumed to be private. In Unity, any public variables are also visible (serialized) in the Inspector window when the script is attached to a GameObject as a component. Ordinarily, private variables are not visible. Using the `[SerializeField]` directive enables a private variable to also be visible and modifiable in the Inspector window.

Congratulations! It's not necessarily a brilliant app, and it's modeled after the example scenes found in the Samples projects, but you started from **File | New Scene** and built it up all on your own. Now, let's have a little fun with it and find a 3D model that's a little more interesting than a little red cube.

Creating a prefab for placing

The prefab object we've been placing on the planes in this chapter is the one named *AR Placed Cube*, which we imported from the AR Foundation Samples project. Let's find a different, more interesting, model to use instead. In the process, we'll learn a bit more about GameObjects, Transforms, and prefabs.

Understanding GameObjects and Transforms

I think a good place to start is by taking a closer look at the *AR Placed Cube* prefab we've been using. Let's open it in the Editor by performing the following steps:

1. In the **Project** window, navigate to the `ARF-samples/Prefabs/` folder.

2. Double-click the **AR Placed Cube** prefab.

We are now editing the prefab, as shown in the following screenshot (I have rearranged my windows differently from the default layout):

Figure 2.10 – Editing the AR Placed Cube prefab

The **Scene** window now shows the isolated prefab object, and the **Hierarchy** window is the hierarchy for just the prefab itself. At its root is an "empty" GameObject named **AR Placed Cube**; it has only one component – Transform, which is required of all GameObjects. Its Transform is reset to **Position** (0, 0, 0), **Rotation** (0, 0, 0), and **Scale** (1, 1, 1).

Beneath the AR Placed Cube is a child **Cube** object, as depicted in the preceding screenshot. This cube is scaled to (0.05, 0.05, 0.05). These units are in meters (0.05 meters is about 2 inches per side). And that's its size when it's placed in the physical environment with our app.

You'll also notice that the child Cube's X-Y-Z **Position** is (0, 0.025, 0), where Y in Unity is the up-axis. As 0.025 is half of 0.05, we've raised the cube half its height above the zero X-Z plane.

The origin of a Cube is its center. So, the origin of the AR Placed Cube is the bottom of the child Cube. In other words, when we place this prefab in the scene, the cube's bottom side rests on the pose position, as determined by the *hit* raycast.

Parenting a model with an empty GameObject to normalize its scale and adjust its origin is a common pattern in Unity development.

Now, let's find a different model for our app and normalize its Transform as we make it a prefab.

Finding a 3D model

To find a 3D model, feel free to search the internet for a 3D model you like. If you're a 3D artist, you may already have ones of your own. You will want a relatively simple, low-poly model (that is, with not many polygons). Look for files in .FBX or .OBJ format, as they will import into Unity without conversion.

I found a model of a virus microbe on cgtrader.com here: https://www.cgtrader.com/free-3d-models/science/medical/microbe. It is a free download and royalty-free, has 960 polygons, and is available in FBX format. My file is named uploads_files_745381_Microbe.fbx.

Once you've found a file and downloaded it to your computer, perform the following steps to import it into Unity:

1. In the **Project** window, create a folder named Models under your _App folder (this step is optional).

2. Drag the model from your Windows File Explorer or macOS Finder into the Models folder to import it into the project. Alternatively, you can use the main menu by clicking **Assets | Import New Asset**.

3. When you select the model in the **Project** window, you can review it in the **Inspector** window. While there, take a look at the many **Import Settings**. Generally, you can keep their default values.

Now, we'll make a prefab of the model and make sure it's been scaled to a usable size. I like to use a temporary Cube object to measure it:

1. In the **Project** window, create a folder named `Prefabs` under your `_App` folder (this step is optional).

2. Right-click inside the `Prefabs` folder, select **Create | Prefab**, and give it a name (I named mine `Virus`).

3. Double-click the new prefab, or click its **Open Prefab** button in the **Inspector** window.

4. For measurement purposes, add a temporary Cube by selecting **GameObject | 3D Object | Cube** from the main menu (or use the + button at the top left, or right-click directly in the **Hierarchy** window).

5. Assuming I want my model to appear in the scene as the same size as the red cube we had been using, set this measuring cube **Scale** to (`0.05,  0.05,  0.05`) and its **Position** to (`0,  0.025,  0`).

6. Drag the 3D model you imported from your **Project** `Models` folder into the **Hierarchy** window as a child of the root object.

7. Use the **Scene** edit toolbar and gizmos to scale and position your model so that it's about the same size and position as the Cube. I found this works: **Scale** (`0.5,  0.05,  0.05`), **Position** (`0,  0.04,  0`), **Rotation** (`0,  0,  0`).

8. Delete or disable the Cube. With **Cube** selected, in its **Inspector** window, uncheck the **Enable** checkbox at the top left.

9. Save the prefab by clicking the **Save** button at the top of the **Scene** window.

The model I found did not come with a material, so let's create one for it now. With the prefab we're working on still open for editing, perform the following additional steps:

1. In the **Project** window, create a folder named `Materials` under your `_App` folder (this step is optional).

2. Right-click inside the `Materials` folder, select **Create | Material**, and give it a name. I named mine `Virus Material`.

3. Drag **Virus Material** onto the model object (**uploads_files_745381_Microbe**) in the **Hierarchy** window.

4. With the microbe model selected in the **Hierarchy** window, you can modify its material in the **Inspector** window. For example, you can change its color by clicking the **Base Map** color chip and choosing a new one. I'll also make mine shinier by setting its **Metallic Map** value to `0.5`.

5. Again, **Save** your prefab.

6. Exit back to scene editing by clicking the < button at the top left of the **Hierarchy** window.

My prefab now looks like this while open for editing (I have rearranged my windows so that they're different from the default layout):

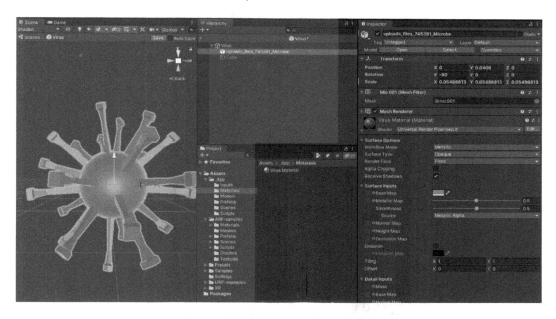

Figure 2.11 – Editing my Virus prefab

We're now ready to add this prefab to the scene. After, we will build and run the finished project.

Completing the scene

We now have our own prefab to place in the AR scene. Let's add it to the **Place Object On Plane** component, as follows:

1. Ensure you've exited the prefab edit mode and are now editing BasicARScene.

2. Select the **AR Session Origin** object in the **Hierarchy** window.

3. From the **Project** window, drag your prefab (mine is `_App/Prefabs/Virus`) into the **Inspector** window, onto the **Place Object On Plane | Placed Prefab** slot.

4. Save the scene with **File | Save**.

5. Build and run the scene by going to **File | Build And Run**.

As shown in the following screenshot, I have infected my desk with a virus!

Figure 2.12 – Running the project shows a virus on my keyboard

There it is. You've successfully created an augmented reality scene that places a virtual 3D model in the real world. Perhaps you wouldn't have chosen a virus, but it's a sign of the times!

You're now ready to proceed with creating your own AR projects in Unity.

Summary

In this chapter, we examined the core structure of an augmented reality scene using AR Foundation. We started with the AR Foundation Samples project from Unity, building it to run on your device, and then exported its assets into an asset package for reuse. Then, we imported these sample assets into our own project, took a closer look at the SimpleAR scene, and built that to run on your device.

Then, starting from a new empty scene, we built our own basic AR demo from scratch that lets the user place a virtual 3D object in the physical world environment. For this, we added **AR Session** and **AR Session Origin** game objects and added components for tracking and visualizing planes and point clouds. Next, we added user interaction, first by creating an **Input Action** controller that responds to screen touches, and then by writing a C# script to receive the OnPlaceObject action message. This function performs a raycast from the screen touch position to find a pose point on a trackable horizontal plane. It then instantiates an object on the plane at that location. We concluded this chapter by finding a 3D model on the internet, importing it into the project, creating a scaled prefab from the model, and using it as the virtual object placed into the scene. Several times along the way, we did a **Build And Run** of the project to verify that our work at that point runs as expected on the target device.

In the next chapter, we will look at tools and practices to facilitate developing and troubleshooting AR projects, which will help improve the developer workflow, before moving on to creating more complete projects in subsequent chapters.

3
Improving the Developer Workflow

When developing for **Augmented Reality** (**AR**), like any software development, it's important to understand your tools, learn how to troubleshoot when you get "stuck," and endeavor to make your overall developer workflow more efficient. In this chapter, we will consider some best practices, techniques, and advanced tools for troubleshooting and testing AR applications in development.

Unity is generally quite friendly for developing for mobile devices. For example, you will normally use the Editor Play-mode to preview your scene in the Editor, allowing rapid *develop-test-update-repeat* cycles. And with an *editor remote tool*, you can run and test on your target mobile device without having to do builds each time.

But Augmented Reality imposes unique challenges because it requires sensor input on the remote device, including a live camera feed and motion sensors. It also requires AR processing built into the mobile software (Android, iOS) that detects features in the environment (such as planes or faces) and tracks your physical device in the real world. Your app requires this data, but it's remote and not normally available to Unity in the Editor Play mode. In this chapter, we'll explore various techniques and tools to deal with this and improve AR development workflows.

In this chapter, you will learn about the following:

- Troubleshooting with log messages
- Debugging with a debugger
- Testing with an editor remote tool
- Simulating environments with the Unity project MARS

If you're impatient and want to begin developing an AR project right away, you may skip this chapter and jump into *Chapter 4*, *Creating an AR User Framework*, where we start our first real project. If that's the case, go ahead but please plan to come back here as soon as you realize this chapter can help you.

Technical requirements

This chapter does not have special technical requirements other than a working development system with Unity installed, a project set up with the XR Plugin and the AR Foundation package, and the ability to successfully build and run on your target device, as given in *Chapter 1*, *Setting Up for AR Development*. The scripts and assets created in this chapter can be found in this book's GitHub repository: `https://github.com/ PacktPublishing/Augmented-Reality-with-Unity-AR-Foundation`.

Troubleshooting with log messages

If (and when) an error occurs while developing or running your Unity project, the first thing you must do is consult the **Console** window for messages. The **Console** window is where you'll find all kinds of messages including asset import warnings, compiler errors, runtime errors in play mode, build problems when you **Build And Run**, and others. Compiler errors (such as coding syntax errors) may prevent the scene from running at all (and the **Play** button will become disabled).

There are three levels of console messages: **Info**, **Warning** (shown in orange), and **Error** (shown in red). You can filter the messages using the toggle buttons in the **Console** window toolbar, as highlighted in the following screen capture:

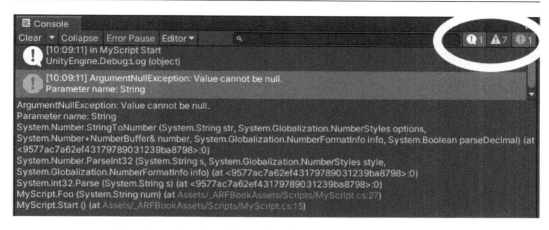

Figure 3.1 – The Console window showing a null exception error

Runtime errors, such as the **ArgumentNullException** error shown in the preceding screenshot, can occur during program execution when you try to use a variable that is not set to any value (more precisely, when its value is `null`).

> **Tip: "Warning" messages can be extraneous**
>
> I generally ignore the warning messages in the Unity Editor's **Console** window, unless I'm deliberately looking for something, as they're often verbose, not relevant to my own problem solving, and thus become noise instead of information. You can hide warning messages by un-clicking the **Warning** button in the Console toolbar.

In the **Console** window, you can often click on an error message to see more details. You might see a list of messages detailing the **stack trace** that provides the runtime state of the program when the error occurred. A stack trace shows the filenames and line numbers in components and Unity code. Although sometimes insightful, you generally will not be interested in traces that refer to built-in Unity or system code. Rather, you'll want to focus on your own scripts. So, look for messages that reference scripts located in your own `Assets/` folder. In the preceding screenshot, *Figure 3.1*, the null exception error occurred on line 27 of the `MyScript.cs` file when it was calling the C# `System.Int32.Parse` function.

> **Tip: Read your Console messages carefully**
>
> A common mistake I often see by novice and experienced developers alike is not reading error messages carefully. When you're in the flow of things, it's often too easy to assume you know what the message is saying and not really read it, missing key clues needed for troubleshooting.

You can also write a message to the Console from your own scripts, using `Debug.Log` calls.

Using Debug.Log

When writing C# scripts, you can log your own messages to the Console using `Debug.Log()` function calls. This is the most common method of checking and understanding what is going on inside your code when it is running. `Debug.Log` messages appear as *Info* messages in the Console (you can also call `Debug.LogError()` to have them appear as *Error* messages instead).

For example, suppose I'm trying to locate the root cause of an error in my project. And suppose there are several `MonoBehaviour` scripts that I'm developing related to this problem. I may place log statements at the entry of specific functions and other log statements to print out specific variables that I am suspicious of. Take the following code, for example, for a script named `MyScript.cs`:

```csharp
// MyScript.cs
using UnityEngine;

class MyScript : MonoBehaviour
{
    public int number;

    void Start()
    {
        number = 10;
    }

    void Update()
    {
        if (number >= 0)
        {
            Debug.Log("in MyScript Update, count = " +
                      number);
            DoSomething();
            number -= 1; // reduce number by one
        }
    }
}
```

```
private void DoSomething()
{
    Debug.Log("inside DoSomething");
    number = -1; // accidently set number to minus-1
    // other code...
}
}
```

In C# you can combine (concatenate) text strings using the plus (+) operator. In our example, the integer `number` is concatenated to the message string (in `Update`), and C# automatically converts the number to a string value first.

Add this script to your scene by creating an empty GameObject (**GameObject | Create Empty**) and dragging the script file from the **Project** window onto the GameObject. Then click **Play**.

When this code runs, what I see in the **Console** window is shown in the following screen capture:

Figure 3.2 – Console messages about my Debug.Log statements

This will reveal that `DoSomething` is only called once rather than 10 times as expected. Can you figure out why?

Studying the code in `Update` does not explain why `DoSomething` was only called once. From there I can re-examine the logic to determine why and when `number` prematurely becomes less than zero. You can see the bug is in the `DoSomething` function itself where it "accidentally" sets `number = -1`, causing the condition in `Update` to never call `DoSomething` after the first time. You may have been stumped while fixated on the `Update` code, but then discovered the bug actually occurs deeper in the program.

> **Tip: Bug hunting? It's probably not where you're looking**
>
> Here's a funny story. A man leaves a bar and sees a drunk guy walking around in circles near a lamppost, searching the sidewalk. "Hey pal, what's the matter?". The drunk replies, "I lost my keys." So together they keep looking. Finally, the man asks, "Are you sure you dropped them here?". The other responds, "Well, I dropped them over there. But the light's better here." Keep this in mind when you're trying to find a bug – it's often exactly *not* where you're looking, otherwise you probably would have found it already!

So far, we've been using the **Console** window to log messages using the Unity Editor play mode. In fact, the Console is so useful for troubleshooting, you may also want to see your debug messages while running your project on your remote device. Next, let's consider how you can use the Console while running on a mobile device connected via USB.

Using the Console with a mobile device

You can use Console logs while running your app on your mobile device, provided the app was built with **Development Mode** enabled, and the device is attached to the Unity Editor via a USB cable (or equivalent). To set this up, use the following steps:

1. Open the **Build Settings** window using **File | Build Settings**.

2. Check the **Development Mode** checkbox.

3. Click the **Build And Run** button.

4. After the app successfully builds, installs on the device, launches, and starts to run, any Debug.Log calls will appear in the Console if you attach it to the application.

   In the Unity **Console** window toolbar, select the **Editor** button and select the process running on your mobile device. For example, the following screenshot shows me attaching the **Editor Console** to my Android device:

Figure 3.3 – Console window attached to an Android device

It's that easy.

There are other kinds of logs provided by Unity and by your device's operating system. In the **Console** window, use the three-dot context menu at the top right to access the full Player logs and Editor logs files. On Android, you can also get more detailed messages from your Android device using *logcat*.

Using logcat with Android devices

On Android mobile devices you can monitor any and all log messages from Android itself and any apps running on the device (including your own Unity one) using a tool called **logcat**. You can install and use *logcat* directly inside the Unity Editor from the Package Manager with the following steps:

1. Use **Window | Package Manager** to open the **Package Manager** window.

2. In the filter select list at the top left, choose **Unity Registry**.

3. Use the search input field at the top right to look for `logcat`.

4. Select the package and click the **Install** button.

5. After the package installs, open the **Logcat** window using **Window | Analysis | Android Logcat**.

With *logcat* installed and its window open, you can run your app on your connected mobile device. It now does not require being built in **Development Mode** enabled nor attached to the Editor Console. You'll discover there is a lot going on inside your device; messages may be streaming from all the running tasks, not just your own application! The **Android Logcat** window offers ways to filter the messages to show only those coming from your app, while your app is running on the device:

- Use the filter drop-down list and choose your app.

- Enter search expressions to filter the message stream.

A screen capture of the **Android Logcat** window is shown in the following figure:

Figure 3.4 – Android Logcat window

I realize the text in this screen capture is probably too small to read here! This screenshot is intended to give you a feel of what the window provides.

> **Info: Using the Android adb command-line tool**
>
> If you are developing for an Android device, I recommend you also install the Android **adb** (Android Debug Bridge) command-line tool. (This is what Unity uses internally for watching Console logs and running the Logcat window.) If you have installed the full *Android Studio* (`https://developer.android.com/studio`), it may already be present on your system. Otherwise, you can install just the command-line tools by navigating to `https://developer.android.com/studio#downloads` and scrolling down the page to the *Command line tools only* section to find the download link for your platform.
>
> With adb installed (and in your command path), you can run a variety of device actions. For more details, see `https://developer.android.com/studio/command-line/adb`. For example, the `adb devices` command will list the Android device it presently sees connected to your computer. `adb logcat` will show the internal device logs. To filter the logs for only Unity-related messages, use the `adb logcat -v time -s Unity` command.

Using the Unity Editor **Console** and `logcat` is great, but what can you do if you want to troubleshoot an app without the mobile device attached to your computer? This can certainly be the case with augmented reality applications that require moving within your environment. One solution is you could create a *virtual* console window, explained next.

> **Info: Using the Xcode console for iOS devices**
>
> If you're developing for iOS, there is no equivalent to logcat in Unity. However, you can view logs from your device using the Xcode log console. Open the console using **View | Debug Area | Activate Console**.

Simulating a Console window in your app

A strategy for capturing console logs, even when your device is detached from the development computer, is to provide a "virtual console" window in your application. This window would be for development, not production. The idea is to replace `Debug.Log` calls with a wrapper function, which optionally outputs to a text object on an in-app text area when running a development build.

We will talk about Unity **Canvas** objects and UI components more in *Chapter 4, Creating an AR User Framework*, so I offer the steps here with only limited explanation.

To implement the wrapper function, use the following steps:

1. In the **Project** window, navigate to your `scripts/` folder (create one if you don't have one yet).

2. *Right-click* in the `scripts` folder, select **Create | C# Script**, and name the script `ScreenLog`.

3. Open the script for editing and replace the default code with the following:

```csharp
// ScreenLog.cs
using UnityEngine;
using UnityEngine.UI;


public class ScreenLog : MonoBehaviour
{
    public Text logText;
    public static ScreenLog Instance { get; private set;
}


    void Awake()
    {
        if (!Instance)
            Instance = this;
    }


    private void Start()
    {
        logText.text = "";
    }
```

```
    private void _log(string msg)
    {
        if (logText)
            logText.text += msg + "\n";
    }

    public static void Log(string msg)
    {
        if (Instance)
            Instance._log(msg);
        Debug.Log(msg);
    } }
```

We implement `ScreenLog` as a **singleton** class (using the `public static ScreenLog Instance` variable), ensuring there will only be one instance of `ScreenLog` in the scene and providing the ability to address the `Log` function as a class method. (We'll discuss *class* versus *instance* methods and the *singleton pattern* more in the next chapter.) This way, you're able to call `ScreenLog.Log()` from anywhere in your own code.

Next, we'll add a text window to the app, and toggle its visibility with a *Debug* button in the UI. (As mentioned, we are going to cover the Unity UI in more detail in later chapters.) First, let's assume your AR application will be used on a mobile device in portrait orientation with a screen space Canvas to contain the text area:

1. From the main menu, create a new Canvas by selecting **GameObject | UI | Canvas**, and rename it `Debug Canvas`. This will also add an **Event System** game object to the scene if one is not already present.

2. To edit the *Screen Space* Canvas, let's switch the **Scene** window to a 2D view by clicking the **2D** button in the **Scene** window toolbar.

3.   It's also helpful to arrange the **Game** window and **Scene** window side by side. Because we're developing for AR, set the **Game** window's display to a fixed portrait aspect ratio, such as **2160x1080 Portrait** using the dimension select list in the **Game** window's top toolbar. This layout can be seen in the following screenshot:

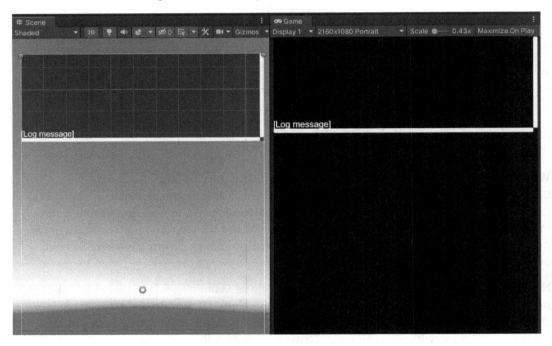

Figure 3.5 – The Scene and Game windows side by side with the portrait device view set up

(This screenshot was captured after all these steps were completed; your Canvas does not have the scrolling text area yet.)

Next, we'll add a scrolling text area where we'll write the log messages:

1.   In the **Hierarchy** window, select **Debug Canvas**. Then right-click, and select **UI | Scroll View**.

2. Resize and place the **Scroll View** in a convenient area of the screen. Use the **Anchor Presets** menu to help align the panel along specific edges of the screen. For example, to anchor it to the top of the screen, select the **Anchor Presets button | Stretch-Top** and then *Shift + Alt* + click **Stretch-Top** and set its **Height** to 400. The resulting **Rect Transform** has **Left, Right**: (0, 0), **Pos Y**: 0, **Anchor Min**: (0, 1), **Anchor Max**: (1, 1), and **Pivot**: (0.5, 1). The **Rect Transform** component and the location of the **Anchor Presets** menu button are shown in the following screenshot:

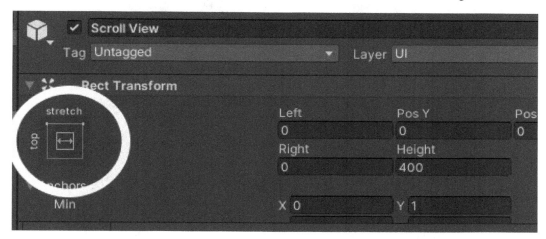

Figure 3.6 – The Rect Transform component with the Anchor Presets button highlighted

3. Allow only vertical scrolling by un-checking the **Scroll Rect | Horizontal** checkbox.

4. Then double-click **Debug Canvas** in **Hierarchy** to bring it into focus (you might need to double-click it twice).

5. In **Hierarchy**, unfold **Scroll View** (the triangle icon) and its **Viewport**. Select the child **Content** object and set its anchors and size by selecting **Rect Transform | Anchor Presets | Stretch-Stretch**, and *Shift + Alt* + click **Stretch-Stretch** to fill the Viewport.

The **Inspector** window for **Scroll View** is shown in the following screenshot:

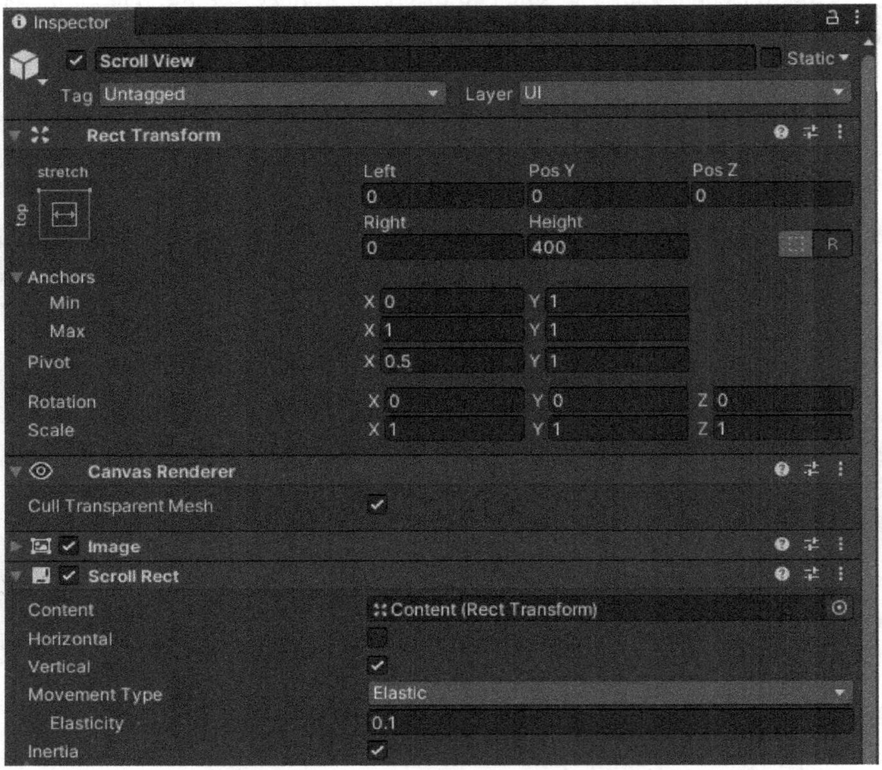

Figure 3.7 – Scroll View property settings

Now we can work on the text element itself:

1. On the **Content** object, *right-click* | **Create** | **UI** | **Text**. Rename it Debug Text. (Note, you may prefer to use *TextMesh Pro* text elements, which give you more control over the typography and padding without any performance costs— introduced in the next chapter.)

2. With the **Debug Text** game object selected, have it fill the **Content** area by using **Anchor Presets** | **Stretch-Stretch**, and *Shift + Alt +* click **Stretch-Stretch**.

3. For reference, enter a placeholder string in the **Text** | **Text** property, such as [Log message].

4. Adjust **Font Size**, for example, to 36.

5. Set **Alignment** to **Bottom**.

6. Change **Vertical Overflow** to **Overflow** (instead of **Truncate**).

7. Personally, I like white text on a black background for the Console. If you agree, in **Scroll View | Image | Color**, set **R, G, B** to 0 and **Alpha** to 200, and in **Debug Text | Text | Color**, set **R, G, B** to 255.

The resulting GameObject **Hierarchy** and **Inspector** settings for the **Debug Text** object are shown in the following screenshot:

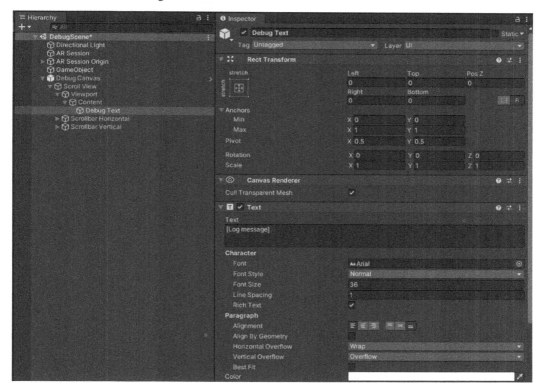

Figure 3.8 – Debug Canvas Hierarchy and Debug Text settings

Next, we will add our script to the scene:

1. With **Debug Canvas** selected in the **Hierarchy** window, locate the **ScreenLog** script in the **Project** window and drag the script file onto the **Debug Canvas** GameObject.

2. With **Debug Canvas** still selected, locate the **Debug Text** GameObject in the **Hierarchy** window, drag it into the **Inspector**, and drop it onto the **Log Text** slot on the **Screen Log** component, as shown in the following screenshot:

Figure 3.9 – Setting the Log Text reference to the Debug Text GameObject

We can now add a Debug button to the UI to toggle the text panel, as follows:

1. *Right-click* on **Debug Canvas** in **Hierarchy** and select **UI | Button** from the menu, to make it a child of the Canvas object. Rename the GameObject to Debug Button.

2. Size and place the button on your screen. For example, set its **Rect Transform | Width, Height** to (175, 175), and anchor it to the lower left of the screen using **Rect Transform | Anchor Presets | Bottom-Left** and *Shift + Alt +* **Bottom-Left**, then set **Pos X, Pos Y** to (30, 30).

3. In **Hierarchy**, unfold **Debug Button** and select its child **Text** object. In the **Inspector**, change its **Text** content to Debug. You can also adjust its font properties from here, such as setting **Font Size** to 36.

4. To change the Debug button into a toggle button, replace the **Button** component with a **Toggle** component.

    In **Hierarchy**, select the **Debug Button** object. In the **Inspector**, use the three-dot context menu icon on the **Button** component (or right-click on the component) and select **Remove Component**.

5. Then click **Add Component**, search toggle, and add a **Toggle** component.

6. Now we'll configure the toggle to handle On Value Changed events. In the **Toggle | On Value Changed** properties, click the small + icon at the bottom right.

7. Drag the **Scroll View** object from the **Hierarchy** window onto the on-click event's **None (Object)** slot. Then, in the **Function** selector, choose **Game Object | Dynamic Bool | SetActive**.

The **Toggle** component now has the following settings:

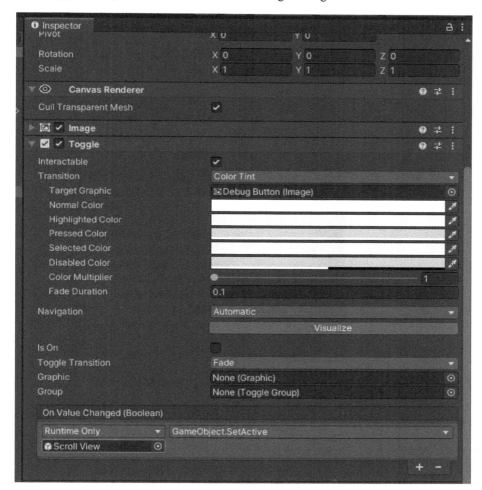

Figure 3.10 – Debug Button set up with a Toggle component

8. Lastly, save this rig as a prefab that you can reuse in other scenes. Drag **Debug Canvas** from the **Hierarchy** window into your Prefabs/ folder in the **Project** window. (If you don't have a Prefabs folder, create one first. The folder name is not required, it's by convention.)

And remember, if you make new changes to this Canvas (or children) in **Hierarchy**, save those changes to the prefab asset using **Overrides | Apply All**.

With this setup, we can now use the `ScreenLog.Log()` function instead of `Debug.Log()` anywhere you want to add an info message in your code, as shown in the following screen capture from my phone:

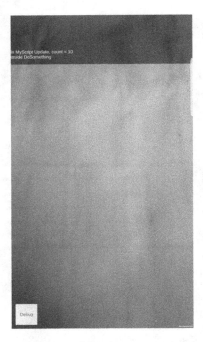

Figure 3.11 – Screenshot of my phone using the virtual console

An advantage of this approach is you can modify it to selectively provide a status message log even for end users, not just your own development.

Info: Third-party virtual consoles

Aside from rolling your own as we do in this chapter, there are third-party virtual console packages you can find in the Asset Store, with a range of features and costs. The *Lunar Mobile Console – Free* asset, for example, is easy to install and use – see `https://assetstore.unity.com/packages/tools/gui/lunar-mobile-console-free-82881`. These tend to be strictly for development purposes and are not appropriate for exposing log messages to end users.

To get a deeper insight into what their code is doing, many programmers like to use a debugger tool provided by **Integrated development environments (IDEs)** such as Visual Studio.

Debugging with a debugger

Professional software developers are familiar with code **debuggers**, used to test and debug programs by stopping the execution at specific lines of code and examining the state of the memory and other runtime conditions. In this section, I will give you an introduction to using the Visual Studio debugger with Unity projects. The debugger can be used in both the Unity Editor play mode, as well as in your builds running on the attached device.

With a debugger, you can set a **breakpoint** at a specific line of code, where the execution will stop at that line, allowing you to query the values of variables, and wait for you to step through or continue the execution of the program.

To use a debugger in the Editor play mode you do not need to make any special changes, provided you are already using Visual Studio for your code editor (or another supported **interactive development environment (IDE)** such as **VS Code** or **JetBrains Rider**). You can configure Unity for your preferred editor/debugger using **Edit | Preferences | External Tools**. For example, in the following screenshot, you can see my Unity install has **External Script Editor** set to a **Visual Studio Community** version:

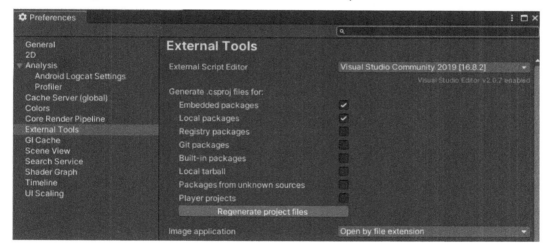

Figure 3.12 – You can set your default code editor in the Unity Preferences window

With Visual Studio opened for your project (choose **Assets | Open C# Project**), you can do more than edit your scripts. You can debug them too. For example, you can set a breakpoint by clicking in the left margin on the line you would like to debug. The following screen capture shows the `MyScript.cs` script is open, and I've created a breakpoint on line 25, indicated in VS Code by a red dot on the screen:

Figure 3.13 – Setting a breakpoint in Visual Studio

To attach the debugger to your Unity Editor session, use the **Attach To Unity** button in the top toolbar. Back in Unity, if you have not yet enabled C# debugging, you will get a prompt like the following:

Figure 3.14 – Unity prompt to enable debugging

Click one of the **Enable** buttons.

Note that Debug Mode can be toggled using the corresponding icon in the bottom-right corner of the Editor window.

Once debugging is enabled and you click **Play** in the Editor, if and when a breakpoint line is reached in your code, execution will stop, and Visual Studio will be given focus on your desktop. The current line of code will be highlighted in yellow on your screen:

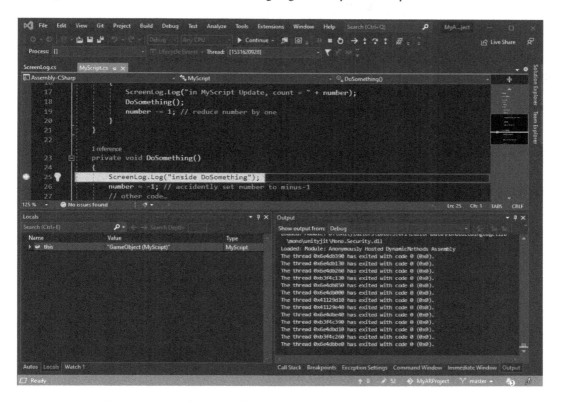

Figure 3.15 – Debugging a line of code in the Visual Studio debugger

There are also debugging windows where you can examine the current values of variables in the script, the current call stack, and so on.

While debugging, the *debugger toolbar* is also active at the top of the window, depicted in the following screenshot:

Figure 3.16 – The debugger toolbar

The **Continue** button (1) will continue running from here until it reaches another breakpoint. The **Stop** button (2) will disable the debugger mode, and the step buttons (3) do the following: **Step Into** follows the code into the body of a function call, **Step Over** will run to the next line of code in the current file, and **Step Up** takes you up the call stack one level.

You can also run the debugger on code running on your attached mobile device.

Debugging on a remote device

To run the debugger on your project running on your mobile device, you must first enable **Script Debugging** and **Development Build** in the project's **Build Settings**. Use the following steps:

1. Open the **Build Settings** window using **File | Build Settings**.
2. Check the **Development Build** checkbox.
3. Check the **Script Debugging** checkbox.
4. Optionally, check the **Wait For Managed Debugger** checkbox.
5. When you're ready, click **Build And Run**.

When the app is running on the device, attach your debugger to the remote process as follows:

1. In Visual Studio, select from the main menu **Debug | Attach Unity Debugger**.
2. A dialog box will appear with a list of potential processes, as depicted in the following screenshot. Choose the process that you want to attach and click **OK**:

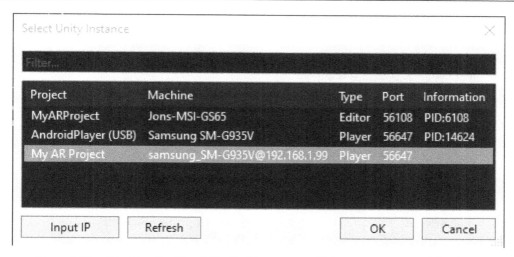

Figure 3.17 – Attaching the Visual Studio debugger to a Unity process on a mobile phone

You can now set and examine breakpoints in the app running on your device. Note, once you close the app on your phone, the debugger also stops in Visual Studio and detaches.

The **Wait For Managed Debugger** build option is useful if you need to start the debugger before Unity starts running. Since Visual Studio needs a process to attach to, the app will start up, then wait for you to attach the debugger, as shown in the following screenshot:

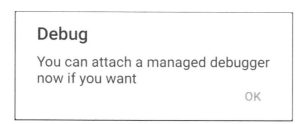

Figure 3.18 – Prompt on phone waiting for a debugger to be attached

In fact, we need this in our little example because the Update function in MyScript will likely be called before I get a chance to attach the debugger.

> **Info: Additional Unity debugging tools**
>
> Unity provides more windows and tools you can use to debug your projects and gain insight into what is going on under the hood. For troubleshooting Input System actions, see **Window | Analysis | Input Debugger**. For deep analysis and profiling, there is the **Profiler, Frame Debugger**, and **Physics Debugger** also under the **Window | Analysis** menu. For the UI, there's the **Immediate Mode GUI (IMGUI)** debugger at **Window | Analysis | IMGUI Debugger**, and when customizing the Unity Editor user interface, see the **Window | UI Toolkit | Debugger** (UI Toolkit is expected to be extended for use in your own apps in the future). There's even a **Window | Render Pipeline | Render Pipeline Debug** window.

Wouldn't it also be good if you could click **Play** to run the project on your mobile device without having to **Build And Run** every time? Let's look at *editor remote tools* next.

Testing with an editor remote tool

Developers have been using Unity for many years to develop games and applications for iOS and Android devices. You want the ability to click **Play** in the Unity Editor and run the current scene remotely on your attached mobile device. Having an iterative *develop-test-update-repeat* cycle is key to more efficient and effective development.

To facilitate this developer workflow, Unity provides an application called **Unity Remote 5** that you install on your phone and then connect to the Unity Editor. It is available for both Android (`https://play.google.com/store/apps/details?id=com.unity3d.mobileremote`) and iOS (`https://apps.apple.com/us/app/unity-remote-4/id871767552`). It allows you to use a mobile device to view and test your project live, inside the Unity Editor, without having to build each time. The device acts as a "remote control" for the scene running in the Editor Play-mode, including screen touch, accelerometer, gyroscope, and webcam input.

Unfortunately, Remote 5 is not suitable for AR development. A remote tool from Unity compatible with AR Foundation has been long promised and is expected, but as I am writing this, it does not exist. Perhaps it will be available by the time you are reading this as a free core Unity feature, so try searching the Unity Forums (`https://forum.unity.com/?gq=AR%20Foundation%20Editor%20Play%20Mode`).

As is often the case in large developer communities, at least one talented individual has stepped up and produced a remote tool for AR Foundation, available on the Unity Asset Store. The *AR Foundation Editor Remote* tool by Kyrylo Kuzyk can be found at `https://assetstore.unity.com/packages/tools/utilities/ar-foundation-editor-remote-168773`. It is not free and, at the present time, it does not support the new Input System, only the legacy Input Manager.

If you choose to purchase the package, you can install it using the Package Manager as follows:

1. Open **Package Manager** using **Window | Package Manager**.

2. Filter the list for **My Assets** using the select list at the top left of the window.

3. Find the **AR Foundation Editor Remote** package, click **Download** (if necessary), then click **Import**. And then in the **Import** dialog box, click the **Import** button.

4. The package is installed in the `Plugins/ARFoundationRemoteInstaller/` folder. The installer should run automatically. Note there is a `Documentation` file as well.

To use the AR Foundation Editor Remote tool, take the following steps as outlined in the documentation:

1. Go to **Edit | Project Settings | XR Plug-in Management |** the **Desktop** tab.

2. Check the **AR Foundation Remote** checkbox.

3. Ensure your project is targeting your mobile device platform in **File | Build Settings**.

4. In the **Project** window, navigate to `Plugins/`
    `ARFoundationRemoteInstaller`, select the **Installer** asset, and view the
    Inspector window as shown in the following screenshot:

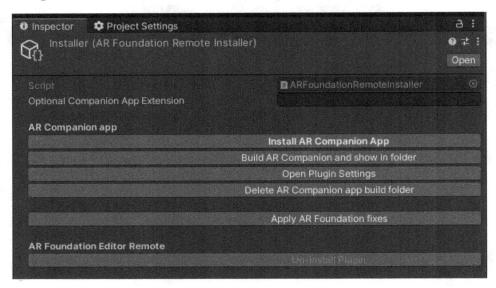

Figure 3.19 – The AR Foundation Remote Installer

5. Click the **Install AR Companion App** button.

    Let it build and install the companion app on your device.

6. In the **Project** window, navigate to the `Plugins/`
    `ARFoundationRemoteInstaller/Resources/` folder and
    select the **Settings** object.

7. Following the instructions on your phone screen and enter the given IP address in
    the **Settings | AR Companion App IP** field in the **Inspector** window.

You're now set up. When you want to use AR Foundation Remote, ensure the AR
Companion app is running on your phone. Then, click **Play** to run your scene using the
mobile device as a remote.

Using an editor remote tool lets you use the Unity Editor Play-mode with your mobile
device. Camera and other sensing data is input into your **Game** window so you can test in
your real-world environment without having to use **Build And Run**.

What if, instead of playing your app on your mobile device, we inverted this approach
by bringing your real-world environment into the Unity Editor? Unity is pioneering this
innovative approach with Unity MARS.

Simulating with Unity MARS

Unity MARS (an acronym for *Mixed Augmented Reality Studio*) (`https://unity.com/products/unity-mars`) is a product solution from Unity Technologies that solves many of the issues with developing Augmented Reality applications discussed thus far in this chapter, and much more.

What is MARS? With MARS you can author and test complex AR applications within the Unity Editor with runtime logic for a range of target physical world environments.

Consider this one scenario: You are developing an AR application for museum visitors, where they point their mobile device at an exhibit or artwork and the app recognizes it and delivers additional information and infotainment, providing a much-enhanced learning experience. But you are at your desk, in your office, across town, or in a different city. How do you develop and test your app? Rather than making travel plans, you could use MARS to bring the target physical space into the Unity Editor and *develop-test-update-repeat* from the comfort of your own desk.

With MARS you can capture and assemble real-world assets such as locations, objects, and props, then drag and drop them into Unity to test them. It supports tracking planes, images, faces, and many other kinds of semantically meaningful data, or *traits*. The MARS documentation can be found at `https://docs.unity3d.com/Packages/com.unity.mars@1.0/manual/index.html`.

The first step in this museum scenario may be to capture the museum's physical environment sensor readings for use at your workstation. The MARS Companion app, described more in this section, can serve this purpose. Likewise, MARS includes a collection of sample environment templates you can use out of the box. The following image, for example, shows a **Simulation View** of a kitchen along with a **Device View** in the same space:

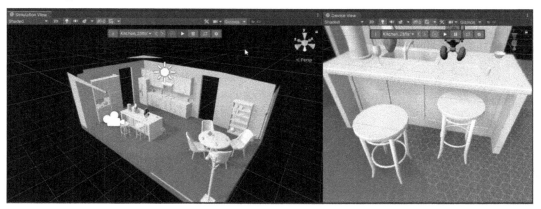

Figure 3.20 – MARS Simulation and Device views

With MARS you have environment simulations that can run either in Edit-mode or Play-mode. You can preview the execution of a scene in Edit-mode, where proxy objects of the real world are copied into the **Simulation** scene view, which is separate from the normal project **Scene** view. You can also start and stop a *continuous* **Simulation** view that is more analogous to the normal Play-mode in Unity. The data fed into the simulation can be synthetic, recorded, or live data. You can then test your project against a wide variety of indoor and outdoor spaces. For a more complete explanation of how MARS simulation works, I recommend this article: `https://blogs.unity3d.com/2020/08/14/a-look-at-how-simulation-works-in-unity-mars`.

MARS provides additional high-level tools and intelligent components that address common challenges for AR developers. Physical environments are not always so predictable. The MARS procedural authoring framework simulates real-world objects, conditions, and actions including "fuzzy authoring" where you specify minimum and maximum measurements for physical features when the app is deciding where and when to let the user interact.

MARS is built on AR Foundation, so it works with all supported AR devices and platforms. Presently, there is a separate annual license fee to use Unity MARS after a trial period.

Using MARS you may still be faced with how to capture your target environment geometry and surface feature for use in simulations. That's where the MARS Companion app comes in.

Capturing with the MARS Companion app

The MARS Companion app can be used to capture real-world data and bring it into the Unity Editor for use with Unity MARS.

Using the app, you can scan a room, take pictures, and record video, capturing and saving this data to the cloud. This data can then be made available to the Unity Editor using the MARS authoring studio.

The app also has limited authoring features that let you create content and layout assets on your device. This could be useful, for example, troubleshooting edge cases where lighting or environment features are ambiguous or difficult to scan.

At this time, the MARS Companion app is still in Beta (`https://forum.unity.com/threads/unity-mars-companion-app-open-beta-announcement.1037638/`) and may eventually be decoupled from MARS for use as an editor remote tool (see the previous section in this chapter).

Unity MARS is a powerful new framework for augmented reality development. It represents Unity's long-term commitment to the AR industry, users, developers, and device manufacturers. Like most Unity packages and modules, it can also be extended with custom behaviors, data extensions, queries, and other add-on modules.

I have not attempted to provide a tutorial for MARS in this chapter (it could be a whole separate book), and *we will not be using MARS in the projects in this book*. Nonetheless, you are more than welcome to improve your workflow using MARS with the projects in this book if you want.

Summary

By its nature, augmented reality mixes the physical and virtual worlds, and that presents unique challenges to AR developers. We develop on a desktop or laptop computer, but the target device for the application is an untethered mobile device. While running an app in Unity Play-mode, an AR scene still needs sensor inputs from the remote device.

In this chapter, we covered a spectrum of tools and techniques that can help with developing and troubleshooting your augmented reality applications using Unity. We started with a basic, classic "print statement," using `Debug.Log()` where you can output log messages to the **Console** window for insight into what is happening in your code. Initially, you might use this just in Play-mode, but we saw how you can build and run your project, and still attach it to the Unity **Console** window to monitor log messages with the app running on your mobile device. Then we built a virtual console window and wrote a `ScreenLog.Log()` wrapper function to optionally let you view log messages on your device without being tethered to Unity at all.

For deeper understanding and to debug your applications, you can use a debugger like the one provided in Visual Studio. While debugging, you can set breakpoints, examine variable values, and step through the code. You can run the debugger both on Unity Play-mode and on applications built and running on your mobile device.

You could also use an editor remote tool—an app that runs on the mobile device and connects to the Unity Editor so you can use the Play-mode and receive input data from the attached device.

Then we took a brief tour of Unity MARS. This AR development studio framework inverts the ordinary remote development paradigm. Rather than running your app on a remote device to capture environment sensor data, MARS lets you use environment sensor simulations directly in the Unity Editor. This provides the opportunity to greatly improve your development workflow and test your application for a wide range of physical environments without leaving your desk.

You are now ready to get started building AR applications. In the next chapter, we develop a framework for controlling user interaction in AR projects. This framework will be saved and used as a template for building and managing the user interfaces in each of the projects in this book.

Section 2 – A Reusable AR User Framework

In this section, we will create a framework for building AR applications with Unity and AR Foundation. Having such a framework generalizes some of the scene structure that I have found myself repeating from one project to the next. The framework manages user interaction modes, user interface panels, and AR onboarding graphics, and we will save it as a template for reuse in other projects in this book. Then, we will show you how to use this framework in a simple place-on-plane project with a main menu.

This section comprises the following chapters:

- *Chapter 4, Creating an AR User Framework*
- *Chapter 5, Using the AR User Framework*

4
Creating an AR User Framework

In this chapter, we will develop a framework for building **Augmented Reality** (**AR**) applications that manage user interaction modes and the corresponding **user interface** (**UI**). The framework includes important **user experience** (**UX**) steps when starting up the AR session at runtime and interacting with AR features. This framework will form the basis for new scenes for projects later in this book.

This is a Unity framework for building mode-based applications. It generalizes some of the scene structure that I have found myself repeating from one project to the next. For example, when an AR app first starts, it must verify that the device supports AR. Once the AR session is initialized, the app may prompt the user to begin scanning the environment to establish tracking. At some point later in the application, the user might be prompted to tap the screen to place a virtual object, often in *Add-object mode*. These steps are common to many AR applications, including the projects in this book, so we will set up some infrastructure beforehand in a scene that may be used as a template.

This chapter involves some advanced C# coding. If you're already an intermediate or advanced programmer, you should be able to follow along fairly easily. If you're a novice, you can just copy/paste the code provided here and learn from it. Or, you have the option of skipping the chapter altogether and using the scene template from this chapter found in this book's GitHub repository.

In this chapter, we will cover the following topics:

- Installing prerequisite assets for our framework
- Starting with a new scene
- Creating the UI canvas and panels
- Creating the UI controller, using a Singleton class
- Creating an interaction modes controller
- Creating the interaction modes, including startup, scan, main, and non-AR modes
- Usng the Unity onboarding UX assets
- Creating a scene template for new scenes

By the end of the chapter, you'll have a scene template, named `ARTemplate`, with AR onboarding features, and a user interaction framework that can be used as a starting point for other AR projects.

Technical requirements

To implement the project in this chapter, you need Unity installed on your development computer, connected to a mobile device that supports AR applications. We'll use the Unity project set up for AR development in *Chapter 1, Setting Up for AR Development*. In review, the project configuration included the following:

- It created a new project (via **Unity Hub**) using the **Universal Render Pipeline** template.
- It set **Target Platform** for **Android** or **iOS** in **Build Settings**, and the corresponding required **Player Settings**.
- It installed an **XR Plugin**, **AR Foundation** package, and configured the **URP Forward Renderer** for AR.
- It installed the **Input System** package and sets **Active Input Handling** (to **Input System Package** or **Both**).

The completed scene from this chapter can be found in this book's GitHub repository at `https://github.com/PacktPublishing/Augmented-Reality-with-Unity-AR-Foundation`.

Understanding AR interaction flow

In an Augmented Reality application, one of the first things the user must do is scan the environment with the device camera, slowly moving their device around until it detects geometry for tracking. This might be horizontal planes (floor, tabletop), vertical planes (walls), a human face, or other objects. A simplistic user flow given in many example scenes is shown in the following diagram:

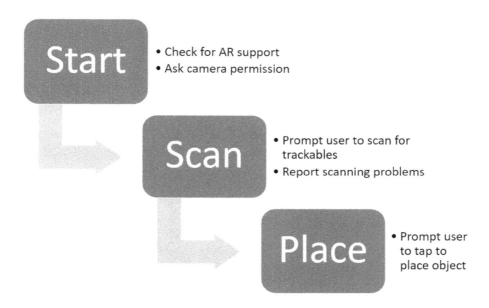

Figure 4.1 – A simple AR onboarding user workflow

As shown in the preceding diagram, the app starts by checking for AR support, asking the user for permission to access the device camera and other initializations. Then, the app asks the user to scan the environment for trackable objects, and may need to report scanning problems, such as if the room is too dark or there's not enough texture to detect features. Once tracking is achieved, the user is prompted to tap the screen to place a virtual object in the scene.

This is great for demo scenes but is probably too simplistic for a real AR application. For example, in the Art Gallery app that we are going to build in *Chapter 6, Gallery: Building an AR App*, after the application starts, the environment is scanned for vertical planes (walls).

Then, the app enters **Main** mode, where the user must tap an **Add** button to add a new picture. That, in turn, displays a modal **Select Image** menu. With pictures added to the scene, the user can pick one and enter **Edit** mode to move, resize, or otherwise modify the virtual object. Part of this general interaction flow is shown in the following diagram:

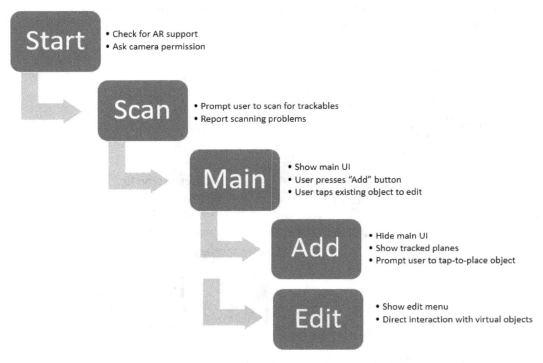

Figure 4.2 – User interaction flow, including Main, Add, and Edit modes

Naturally, each application has its own interaction flows. The framework we are building in this chapter supports this scenario and can be adapted for other projects that require managing a current modal state and corresponding UI.

This framework implements a **state machine** design pattern, where the scene has a current *state* (interaction mode and visible UI). Specific conditions must be met to then transition from one state to another.

There are two major areas of this framework – the UI panels and the interaction modes. Generally, there will be a one-to-one correlation between the modes and the UI used by the modes. For example, in *Main mode*, there will be the main menu UI. In *Add-object mode*, there will be a UI prompt for the user to tap to place an object in the scene. This implements a design pattern called **view-controller**, with UI views and mode controllers.

Let's now begin to implement this basic workflow in our scene by adding a number of additional prerequisite packages to the project.

Installing prerequisite assets

Our user interaction framework uses several additional packages that need to be installed in your project, namely, TextMeshPro, DOTween, and Serialized Dictionary Lite. In this section, I will also include some utility assets. Let's install them now.

TextMeshPro

TextMeshPro provides high-quality text assets that replace the built-in text element. It is not mandatory, but I strongly recommend it. To import **TextMeshPro**, if you haven't installed it yet in your project, perform the following steps:

1. Go to **Window | TextMeshPro | Import TMP Essential Resources**.
2. In the **Import Unity Package** window, click **Import**.

The TextMeshPro package is now installed. You may also install the **TMP Examples and Extras** package, which includes additional fonts and other assets that may be useful and fun for your projects.

DOTween

DOTween is, in my opinion, an indispensable free package for doing small, lightweight animation effects on just about any `MonoBehaviour` property. Without it, you may need to write a dozen lines of code to do what DOTween does in one. Documentation for DOTween can be found online at `http://dotween.demigiant.com/documentation.php`.

To add DOTween, perform the following steps:

1. Go to its Unity Asset Store page: `https://assetstore.unity.com/packages/tools/animation/dotween-hotween-v2-27676`.
2. Press **Add to My Assets** and/or **Open In Unity**.
3. This will take you to the **Package Manager** window in your Unity project.
4. Ensure **My Assets** is selected from the **Packages** filter dropdown in the upper-left corner of the **Package Manager** window.
5. Search for `DOTween` using the search text input field in the upper-right corner of the **Package Manager** window.

6. Select the **DOTween** package and then click **Install**.

7. Once imported, you are prompted to **Open DOTween Utility Panel** to set up the package.

8. Then, click the **Setup DOTween** button.

DOTween is now installed and set up on your project.

Serialized Dictionary Lite

A C# **dictionary** is a key-value list structure where values in the list can be referenced by a key value. For example, we will use dictionaries to look up a UI panel or interaction mode object by name. Unfortunately, Unity does not provide native support for dictionaries in the Editor's **Inspector** window. **Serialized Dictionary Lite** is a free extension to the Unity Editor that allows dictionaries to be edited using **Inspector**. To add Serialized Dictionary Lite to your project, perform the following steps:

1. Go to its Unity Asset Store page, `https://assetstore.unity.com/packages/tools/utilities/serialized-dictionary-lite-110992`.

2. Press **Add to My Assets** and/or **Open In Unity**.

3. This will take you to the **Package Manager** window in your Unity project.

4. Ensure **My Assets** is selected from the **Packages** filter dropdown in the upper-left corner of the **Package Manager** window.

5. Search for `Serialized` using the search text input field in the upper-right corner of the **Package Manager** window.

6. Select the **Serialized Dictionary Lite** package and click **Install** (or, if prompted, click **Download** and then **Import**).

Serialized Dictionary Lite is now installed in your project.

Other prerequisite assets

In addition to the aforementioned packages, we will assume that you have the following already added to your Unity project:

*   Assets from the Unity **arfoundation-samples** project imported from the `ARF-samples.unity` package created in *Chapter 2, Your First AR Scene*.

- In *Chapter 2*, *Your First AR Scene*, we also created an **AR Input Actions** asset containing an **Action Map** named **ARTouchActions**, including (at least) one **PlaceObject** action.

With our prerequisite assets present, we can get started with building the scene.

Starting with a new scene

We start this project with a new empty scene and set it up with the AR Foundation objects: **AR Session** and **AR Session Origin**. Create a new scene named `ARFramework` using the following steps:

1. Create a new scene using **File | New Scene**.
2. Choose the **Basic (Built-in)** template. Press **Create**.
3. Save the scene using **File | Save As**, navigate to your `Assets/Scenes/` folder, give it the name `ARFramework`, and then click **Save**.

Next, we'll set up the scene with the basic AR Foundation game objects as follows:

1. Delete **Main Camera** from the **Hierarchy** window by *right-clicking* and selecting **Delete** (or the pressing *Del* key on your keyboard).
2. Add an **AR** session by selecting **GameObject** from the main menu, and then **XR | AR Session**.
3. Add an **AR Session Origin** object by selecting **GameObject** from the main menu, and then **XR | AR Session Origin**.
4. Select the **AR Session Origin** object in the **Hierarchy** window. In the **Inspector** window, click **Add Component**, search for `raycast`, and then add an **AR Raycast Manager** component.
5. Unfold **AR Session Origin** and select its child **AR Camera**. In the **Inspector** window, use the **Tag** selector in the upper-left corner to set its tag to **MainCamera**. (This is not required, but it is a good practice to have one camera in the scene tagged as MainCamera).
6. In the **Inspector** window, click **Add Component**, search for `audio listener`, and add an **Audio Listener** component to the camera.

For demo purposes, we'll add an **AR Plane Manager** component for detecting and tracking horizontal planes. This may change based on the requirements of a specific project:

1. With **AR Session Origin** selected in the **Hierarchy** window, click **Add Component** in the **Inspector** window, search for `ar plane manager`, and then add an **AR Plane Manager** component.

2. Choose an AR plane visualizer prefab and add it to the **Plane Prefab** slot. For example, try the **AR Plane Debug Visualizer** prefab found in the `ARF-samples/Prefabs` folder.

We can also set up some basic AR light estimation as follows:

1. Select **Main Camera** in the **Hierarchy** window. On its **AR Camera Manager** component, set **Light Estimation** to **Everything**.

2. In the **Hierarchy** window, select the **Directional Light** game object. In the **Inspector** window, click **Add Component**, search for `light estimation`, and then add a **Basic Light Estimation** component.

3. Drag the **AR Camera** object from the **Hierarchy** window onto the **Basic Light Estimation | Camera Manager** slot.

4. Save your work using **File | Save**.

We now have a scene named `ARFramework` with a few things set up, including the AR Session, AR Session Origin, AR Camera, and basic light estimation. We can now begin to construct our framework's UI panels.

Creating the UI canvas and panels

The main screen space UI canvas will contain various user interface panels that may be displayed at various times throughout the application. Presently, we'll include the following UI panels.

- The Startup UI panel with any initialization messages

- The Scan UI panel, which prompts the user to scan for trackable features

- The Main UI panel for the main mode that could display the main menu buttons

- The NonAR UI panel, which could be shown when the device does not support Augmented Reality

Creating the screen space canvas

First, we need to create a Canvas to contain these panels. Follow these steps:

1. From the main menu, select **GameObject | UI | Canvas** and rename the Canvas UI Canvas. We can leave the default **Render Mode** as **Screen Space – Overlay**. This will also add an **Event System** game object to the scene if one is not already present.

2. By default, the new Canvas is in screen space, and this is what we want here. Some people prefer to change **Canvas Scaler UI Scale Mode** from **Constant Pixel Size** to **Scale With Screen Size**.

3. To edit a Screen Space canvas, let's switch the **Scene** window to a 2D view by clicking the **2D** button in the **Scene** window toolbar. Then, double-click the **UI Canvas** object in the **Hierarchy** window to focus the **Scene** view on this object.

4. It's also helpful to arrange the **Game** window and **Scene** window side by side. Because we're developing for AR, set the **Game** window's display to a fixed portrait aspect ratio, such as **2160x1080 Portrait** using the dimension select list in the **Game** window's top toolbar.

On this canvas, we will add the separate panels. First, let's add an app title at the top of the screen.

Adding an app title

Let's add a placeholder for an app title as a text panel positioned at the top of the screen. Add the title using the following steps:

1. *Right-click* on **UI Canvas** and select **UI | Panel**. Rename the panel App Title Panel.

2. With the **App Title Panel** object selected, in its **Inspector** window, open the **Anchor Presets** menu (found in the upper-left corner of the **Rect Transform** component), and click the **Stretch-Top** button. The **Anchor Presets** menu is shown open in the following screenshot, to the left of the **Rect Transform** component:

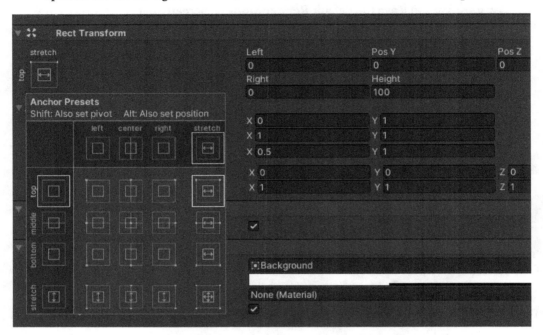

Figure 4.3 – Anchor Presets menu for App Title Panel set to Top-Stretch

3. Then, press *Shift* + *Alt* + **Stretch-Top** to set its pivot and position.

4. Set **Rect Transform | Height** to 100.

5. Next, *right-click* on **App Title Panel**, select **UI | Text – TextMeshPro**, and rename the object Title Text.

6. In its **TextMeshPro – Text** component, set **Text Input** to My AR Project.

7. Using the **Anchor Presets** menu in the upper-left corner of **Rect Transform**, select **Stretch-Stretch**. Then, press *Shift* + *Alt* + **Stretch-Stretch**.

8. Set **Alignment** to **Center** and **Middle**.

9. You may also choose to adjust the **Font Size** and **Vertex Color** fields as you wish.

There isn't much to see, but the **Game** window, along with the title of the app, is shown in the following screenshot:

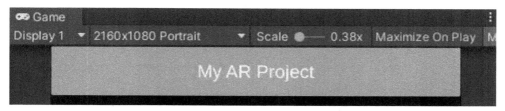

Figure 4.4 – Game window (cropped) with the App Title panel anchored as Top-Stretch

Now that you have experience using the **Anchor Presets** menu, I'll abbreviate the instructions going forward. Next, we'll add a panel for the start up mode.

Creating the UI panels

We'll now create the UI panels for each of the initial interaction modes supported by the framework. Since they are all very similar, we'll create the first one, and then duplicate and modify it for the others.

The first UI panel, **Startup UI**, will be a text panel displayed when the app is initializing. Create it using the following steps:

1. In the **Hierarchy** window, *right-click* the **UI Canvas** object and select **UI | Panel**. Rename it `Startup UI`.

2. We don't need a background image so, in the **Inspector** window, remove the **Image** component using the *3-dot context menu* | **Remove Component**.

3. Click the **Add Component** button, search for `canvas group`, and add a **Canvas Group** component to the panel. We're going to use this component to fade panels on and off later in this chapter.

4. *Right-click* the **Startup UI** object and select **UI | Text – TextMeshPro**.

5. Set **Text Input** to `Initializing…`.

6. Using its **Anchor Presets** menu, select **Stretch-Stretch**. Then, press *Shift + Alt +* **Stretch-Stretch**.

7. Set **Alignment** to **Center** and **Middle.**

Next, we can add a panel that can be displayed if the device we're running on does not support AR. Create this panel as follows:

1. *Right-click* the **Startup UI** panel and select **Duplicate**. Rename it to `NonAR UI`.

2. Unfold the object and select its child text object. Change the text content to `Augmented reality not supported on this device.`

The Scan UI panel will be used to prompt the user to scan the room while the app tries to detect AR features. Create the panel by following these steps:

1. *Right-click* the **Startup UI** panel and select **Duplicate**. Rename it to `Scan UI`.

2. Unfold the object and select its child text object. Change the text content to `Scanning... Please move device slowly`.

Lastly, we'll add a placeholder panel for the main mode UI. This panel could later include, for example, a main menu for the app:

1. *Right-click* the **Startup UI** panel and select **Duplicate**. Rename it to `Main UI`.

2. Unfold the object and select its child text object. For development purposes, change the text content to `Main Mode Running`.

The current UI Canvas hierarchy is shown in the following screenshot:

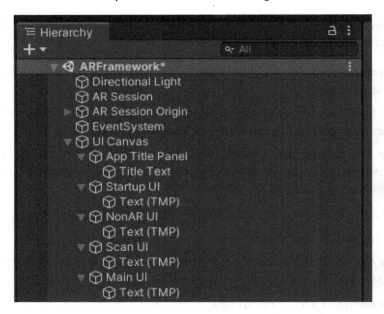

Figure 4.5 – UI Canvas hierarchy

So far, we have created a simple hierarchy of UI panels under a screen space UI Canvas. The panels are acting as a placeholder, for the most part, containing a text element so that you can see which panel is active at runtime. As you build your own apps from this scene, you'll fill in the panels with app-specific UI elements.

Next, we'll create the UI controller script.

Creating the UI controller

It will be convenient to have a script with a small API that makes it easy to switch between UI panels. For the controller scripts in our framework, I've decided to define them as singletons.

A **singleton** is a software design pattern that ensures there is only a single instance of a script object at runtime. Then, the object's instance can be more easily referenced, using a static reference to `Instance` in the class definition. Learn more at `https://wiki.unity3d.com/index.php/Singleton`.

Then, we'll write a `UIController` script that controls the visibility of your UI panels. Lastly, we'll implement some code to fade in and out for a more pleasing user experience when we hide and show the panels.

Creating a Singleton class script

We'll begin by writing a `Singleton` class to use (or, if you already have a favorite, feel free to use that `Singleton` class definition instead). You can find some singleton implementations available as packages in the Unity Asset Store, but all we need is a short script that you can now create as follows:

1. In your **Project** window, create a new C# script in your `Scripts/` folder by *right-clicking* and selecting **Create | C# Script**, and name it `Singleton`.

2. Write the script as follows:

```
using UnityEngine;


///     Singleton behaviour class, used for components
        that should only have one instance
/// </summary>
/// <typeparam name="T"></typeparam>
public class Singleton<T> : MonoBehaviour where T :
Singleton<T>
{
    public static T Instance { get; private set; }


    /// <summary>
    ///     Returns whether the instance has been
            initialized or not.
    /// </summary>
    public static bool IsInitialized {
```

```
            get { return Instance != null; }
    }


    /// <summary>
    ///      Base awake method that sets the singleton's
            unique instance.
    /// </summary>
    protected virtual void Awake()
    {
        if (Instance != null)
            Debug.LogError($"Trying to instantiate a
                second instance of singleton class
                    {GetType().Name}");
        else
            Instance = (T)this;
    }


    protected virtual void OnDestroy()
    {
        if (Instance == this)
            Instance = null;
    }
}
```

3. Save the file.

> **Info: A singleton as an anti-pattern**
>
> Note that the singleton pattern can be abused, and some programmers are adamantly opposed to using it, as it can cause problems down the road should your application grow and get more complex. But it's a powerful tool when you are certain that the app will only ever require one instance of the class, as will be the case in this interaction framework. One of the main advantages of singletons is that you can then reference the object instance as a static variable on the object class itself. An alternative technique is to find the instance to the component at runtime, for example, by calling `FindObjectOfType<T>()` from the script's `Start()` function.

This script can be used to declare a singleton's `MonoBehaviour` class, as we'll see next in `UIController` and other scripts.

Writing the UIController script

With our Singleton class in hand, we can now write a UI controller. This component provides a way to switch between UI panels visible to the user. Perform the following steps to write the `UIController` class:

1. Begin by creating a new script in your **Project** `Scripts/` folder by *right-clicking and selecting* **Create | C# Script.** Name the script `UIController`.

2. *Double-click* the file to open it for editing and replace the default content, starting with the following declarations:

```
using UnityEngine;
using RotaryHeart.Lib.SerializableDictionary;

[System.Serializable]
public class UIPanelDictionary :
SerializableDictionaryBase<string, CanvasGroup> { }

public class UIController : Singleton<UIController>
{
    [SerializeField] UIPanelDictionary uiPanels;

    CanvasGroup currentPanel;
```

At the top, we declare a serializable dictionary, `UIPanelDictionary`, using the **Serializable Dictionary Lite** package's base class (we installed this package as a prerequisite earlier in this chapter. See `https://assetstore.unity.com/packages/tools/utilities/serialized-dictionary-lite-110992` and the associated Unity Forum for documentation). The dictionary lookup key is the UI's name, and its value is a reference to the UI panel's `CanvasGroup` component.

Instead of declaring `UIController` as a `MonoBehaviour` class, we declare it a `Singleton` (which itself derives from `MonoBehaviour`). Don't worry about the syntax of the declaration, `public class UIController : Singleton<UIController>`. This is what our `Singleton` class expects.

The script declares a `uiPanels` variable as a `UIPanelDictionary`. We also declare a `currentPanel` variable to track which panel is presently active.

3.   Next, add the following functions to the script, which ensure all the UI panels are disabled when the app is started, by iterating through the `uiPanels` list and calling `SetActive(false)`:

```
void Awake()
{
    base.Awake();
    ResetAllUI();
}
void ResetAllUI()
{
    foreach (CanvasGroup panel in uiPanels.Values)
    {
        panel.gameObject.SetActive(false);
    }
}
}
```

Note that `Awake` calls `base.Awake()` because the parent `Singleton` class also has an `Awake` that must be called in order for this to work. Then it calls `ResetAllUI`.

4.   Then, add the following functions to the script:

```
public static void ShowUI(string name)
{
    Instance?._ShowUI(name);
}

void _ShowUI(string name)
{
    CanvasGroup panel;
    if (uiPanels.TryGetValue(name, out panel))
    {
        ChangeUI(uiPanels[name]);
    }
    else
    {
        Debug.LogError("Undefined ui panel " + name);
```

```
        }      }

    void ChangeUI(CanvasGroup panel)
    {
        if (panel == currentPanel)
            return;
        if (currentPanel)
            currentPanel.gameObject.SetActive(false);
        currentPanel = panel;
        if (panel)
            panel.gameObject.SetActive(true);
    }
```

_ShowUI is an instance function that, given a panel name, calls ChangeUI. ChangeUI hides the current panel and then activates the required one (note that I'm using an underscore prefix to distinguish private instance functions from the public one). The C# dictionary, TryGetValue, looks up the value for the given key.

The static ShowUI class function simply calls the instance's _ShowUI function. In this way, another script can show a panel by calling UIController. ShowUI(panelname); without requiring a direct reference to the instance. It uses the null-conditional operator (https://docs.microsoft.com/en-us/dotnet/csharp/language-reference/operators/member-access-operators#null-conditional-operators--and-) as a shortcut to make sure the instance is defined before we reference it.

Now, add the script as a component on the UI Canvas and set up its properties by performing the following steps:

1. In the **Hierarchy** window, select **UI Canvas**.

2. Drag the UIController script onto **UI Canvas**, adding it as a component.

3. In the **Inspector** window, on the **UI Controller** component, unfold the **UI Panels** dictionary list.

4. Click the + button in the bottom-right corner of the UI Panels list.

5. In the elements **Id** slot, write Startup.

6. Unfold the element and then, from the **Hierarchy** window, drag the **Startup UI** game object onto the **Value** slot.

7.   Repeat steps 4 – 6 three times for each of the following: NonAR : **NonAR UI**, Scan : **Scan UI**, and Main : **Main UI**.

The UI Controller component should now look like the following:

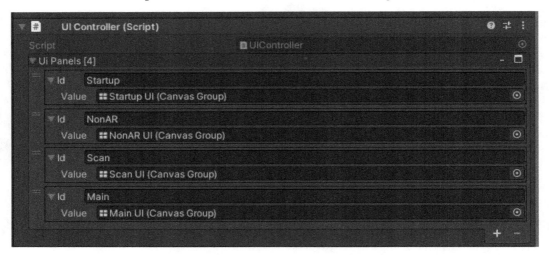

Figure 4.6 – UI Controller component populated with UI panel references

Thus far, we have created a simple UI for an AR application, organized on one canvas as a set of separate panels. Our plan is to present only one panel at a time to the user, depending on what the application is doing. We also wrote a UIController script to handle switching between panels.

Fading the UI panels

An improvement we can make is to fade the UI in and out while transitioning instead of abruptly hiding/showing a panel. Presently, we call SetActive to change the panel's visibility. Instead, we can use the panel's CanvasGroup component and animate its Alpha value, and the DOTween library is very handy for this. (You can skip this modification if you do not want to install DOTween). To do this, follow these steps:

1.   Open the UIController script for editing and add the following declaration at the top of the file:

```
using DG.Tweening;
```

2. Add these two fader helper functions at the bottom of the class:

```
void FadeIn(CanvasGroup panel)
{
    panel.gameObject.SetActive(true);
    panel.DOFade(1f, 0.5f);
}
void FadeOut(CanvasGroup panel)
{
    panel.DOFade(0f, 0.5f).OnComplete(() => panel
        gameObject.SetActive(false));
}
```

3. Then, modify the `ChangeUI` function to call the fader helps instead of `SetActive`, as shown here (the lines in comments are replaced):

```
void ChangeUI(CanvasGroup panel)
{
    if (panel == currentPanel)
        return;
    if (currentPanel)
        FadeOut(currentPanel);
        //currentPanel.gameObject.SetActive(false);

    currentPanel = panel;

    if (panel)
        FadeIn(panel);
        //panel.gameObject.SetActive(true);
}
```

Eventually, when you run the scene, the UI panels will fade in and out when shown and hidden, respectively.

Next, we will write an Interaction Controller that handles the application interaction modes and uses the UI Controller to display the specific UI it needs.

Creating an Interaction Controller mode

For our user framework, we will make a clever use GameObject with a mode script on it to represent interaction modes. Modes will be enabled (and disabled) by enabling (and disabling) the corresponding objects. We'll organize these objects in a hierarchy, like the UI panels we created in the previous section, but separated to keep the "controllers" apart from the "views," as prescribed by the controller/view software pattern. Presently, we'll include the following modes:

- **Startup mode**: Active while the AR session is initializing, and then it initiates Scan mode.

- **NonAR mode**: A placeholder should you want your application to run even if the device does not support AR.

- **Scan mode**: This prompts the user to scan for trackable features until the AR session is ready, and then it initiates Main mode.

- **Main mode**: This displays the main menu and handles non-modal interactions.

First, we'll create the object hierarchy representing each of these modes, under an Interaction Controller game object. With separate GameObjects representing each mode, we'll be able to enable one mode or another separately.

Creating the interaction mode hierarchy

To create the interaction mode hierarchy, perform the following steps:

1. From the main menu, select **GameObject | Create Empty**, and rename the object `Interaction Controller`.

2. *Right-click* the **Interaction Controller** object and select **Create Empty**. Rename it `Startup Mode`.

3. Repeat *step 2* three more times to create objects named `NonAR Mode`, `Scan Mode`, and `Main Mode`.

The mode hierarchy game objects now look like the following:

Figure 4.7 – Interaction Controller modes hierarchy

Now we can write and set up the `InteractionController` script.

Writing the Interaction Controller

The role of our Interaction Controller is to manage the top-level user interaction of the application. We'll begin by writing the script as follows:

1. Create a new script in your **Project** `Scripts/` folder by *right-clicking* **Create C# Script** and name the script `InteractionController`.

2. *Double-click* the file to open it for editing and replace the default content, starting with the following declarations:

```
using System.Collections;
using UnityEngine;
using RotaryHeart.Lib.SerializableDictionary;


[System.Serializable]
public class InteractionModeDictionary :
SerializableDictionaryBase<string, GameObject> { }


public class InteractionController :
Singleton<InteractionController>
{
    [SerializeField] InteractionModeDictionary
        interactionModes;


    GameObject currentMode;
}
```

At the top, we declare a serializable dictionary, `InteractionModeDictionary`, using the **Serializable Dictionary Lite** package's base class. The dictionary key is the mode's name, and its value is a reference to the mode game object.

Instead of declaring `InteractionController` as a `MonoBehaviour` class, we declare it a `Singleton` (which itself derives from `MonoBehaviour`).

Then we declare the `interactionModes` variable as this type of dictionary. We also declare a `currentMode` variable that tracks the current enabled mode.

3. Next, add the following functions to the script, which ensures all the modes are disabled when the app is started, by iterating through the `interactionModes` list by calling `SetActive(false)`:

```
protected override void Awake()
{
    base.Awake();
    ResetAllModes();
}

void ResetAllModes()
{
    foreach (GameObject mode in interactionModes
        Values)
    {
        mode.SetActive(false);
    }
}
```

Note that `Awake` calls `base.Awake()` because the parent `Singleton` class also has an `Awake` that must be called in order for this to work. It then calls `ResetAllModes`.

4. Then, add the following functions to the script:

```
public static void EnableMode(string name)
{
    Instance?._EnableMode(name);
}

void _EnableMode(string name)
{
    GameObject modeObject;
    if (interactionModes.TryGetValue(name, out
        modeObject))
    {
        StartCoroutine(ChangeMode(modeObject));
    }
    else
```

```
            {
                Debug.LogError("undefined mode named " +
                    name);
            }       }

    IEnumerator ChangeMode(GameObject mode)
    {
        if (mode == currentMode)
            yield break;

        if (currentMode)
        {
            currentMode.SetActive(false);
            yield return null;
        }

        currentMode = mode;
        mode.SetActive(true);
    }
```

_EnableMode is an instance function that, given a mode name, calls ChangeMode. ChangeMode disables the current mode and then activates the requested one.

Note that ChangeMode is called as a **coroutine** to allow the current mode an extra frame to be disabled before activating the new one. (To learn more about coroutines, see https://docs.unity3d.com/Manual/Coroutines. html).

The static EnableMode class function simply calls the instance's _EnableMode function. In this way, another script can show a panel by calling InteractionController.EnableMode(modename); without requiring a direct reference to the instance. It uses the null-conditional operator (https:// docs.microsoft.com/en-us/dotnet/csharp/language-reference/ operators/member-access-operators#null-conditional- operators--and-) as a shortcut to make sure the instance is defined before we reference it.

5. Lastly, assuming we want the app to start in `Startup` mode, add the following:

```
void Start()
{
    _EnableMode("Startup");
}
```

This assumes we will include a `"Startup"` mode in the `interactionModes` dictionary.

`UIController` will contain references to each of the app's mode game objects. When the app needs to switch modes, it will call `InteractionController.EnableMode(modeName)` with the name of the mode. The current mode will be disabled, and the required one will be enabled.

Add the script as a component on the Interaction Controller game object and set up its properties by following these steps:

1. In the **Hierarchy** window, select the **Interaction Controller** game object.
2. Drag the `InteractionController` script onto the **Interaction Controller**, adding it as a component.
3. In the **Inspector** window, on the **Interaction Controller** component, unfold the **Interaction Modes** dictionary list.
4. Click the + button in the bottom-right corner of the **Interaction Modes** list.
5. On the elements **Id** slot, write `Startup`.
6. Unfold the element and then, from the **Hierarchy** window, drag the **Startup Mode** game object onto the **Value** slot.
7. Repeat steps 4 – 6 three times for each of the following: `NonAR` : **NonAR Mode**, `Scan` : **Scan Mode**, and `Main` : **Main Mode**.

The **Interaction Controller** component should now look like the following:

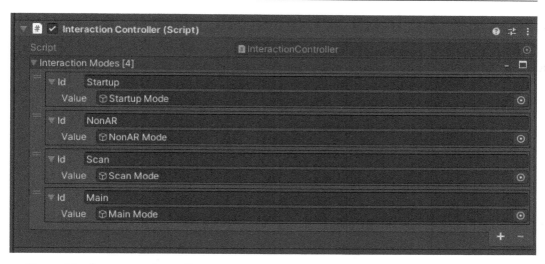

Figure 4.8 – Interaction Controller component populated with interaction mode object references

8. The **Interaction Controller** component will be responding to user input, so we need to add a **Player Input** component (assuming your project is using the new Input system).

    With **Interaction Controller** selected in the **Hierarchy** window, click **Add Component** in the **Inspector** window.

9. Search for `player inp ut` and add a **Player Input** component.

10. Locate the **AR Input Actions** asset in your **Project** window (for example, the `Inputs/` folder) and drag it to the **Player Input | Actions** slot. (As noted in the *Technical requirements* earlier in the chapter, I assume you already have this asset as created in *Chapter 2, Your First AR Scene*).

11. Set **Player Input | Behavior** to **Broadcast Messages**.

    THIS IS IMPORTANT! We need to make sure the player actions are forwarded to the child mode objects.

In this section, we have created a hierarchy for interaction modes, organized under one Interaction Controller game object that has a script for enabling/disabling mode objects. Our plan is to allow only one mode to be active at a time. Of course, we still need to write the scripts that control each mode, and handle conditions when it's time to transition from one particular mode to a different one.

Creating the interaction modes behavior

When the app enables a mode, it will enable the corresponding game object, which has a script that controls the behavior of that mode. When the app changes modes, the current mode object will be disabled, and the new one enabled. Each mode is responsible for the following:

- Displaying its corresponding UI
- Transitioning to a different mode when specific conditions are met

We will write mode scripts for each of the modes.

The StartupMode script

Startup mode begins when the application starts (it's enabled from the `InteractionController Start()` function). It displays the Startup UI panel. Then it waits for the `ARSession` state to become ready, and transitions to Scan mode. Or, if the `ARSession` reports that AR is not supported on the current device, it transitions to NonAR mode.

Follow these steps to create Startup mode:

1. Create a new script in your **Project** `Scripts/` folder by *right-clicking* and selecting **Create | C# Script,** and name the script `StartupMode`.

2. Drag the `StartupMode` script onto the **Startup Mode** game object in the **Hierarchy** window.

3. *Double-click* the `StartupMode` script file to open it for editing and write it as follows:

```csharp
using UnityEngine;
using UnityEngine.XR.ARFoundation;

public class StartupMode : MonoBehaviour
{
    [SerializeField] string nextMode = "Scan";

    void OnEnable()
    {
        UIController.ShowUI("Startup");
    }
```

```
    void Update()
    {
        if (ARSession.state ==
            ARSessionState.Unsupported)
        {
            InteractionController.EnableMode("NonAR");
        }
        else if (ARSession.state >= ARSessionState.Ready)
        {
            InteractionController.EnableMode(nextMode);
        }
    }
}
```

The script uses the AR Foundation's ARSession class state variable, ARSession. state, to determine when the session is initialized or whether AR is unsupported. The state is an enum ARSessionState with one of the following values:

- None: The session has not yet been initialized.
- Unsupported: The device does not support AR.
- CheckingAvailability: The session is in the process of checking availability.
- NeedsInstall: The device needs to install or update AR support software.
- Installing: The device is in the process of installing AR support software.
- Ready: The device supports AR and you can enable the ARSession component.
- SessionInitializing: The AR session is scanning the environment and trying to detect trackable objects.
- SessionTracking: The AR session has found trackable objects and can determine the device's location within the real-world 3D environment.

When state is Unsupported, we transition to NonAR mode.

When state is Ready (or higher), we transition to Scan mode.

The ScanMode script

Scan mode is enabled when the device is scanning the environment, trying to detect trackable features in the real world. It displays a prompt asking the user to point the camera into the room and slowly move the device.

The conditions for ending Scan mode may vary depending on the AR application. For example, it may wait until at least one horizontal or vertical plane has been detected, or a reference image has been recognized, or a selfie face is being tracked. Presently, we'll check ARPlaneManager if any trackables have been detected.

Perform the following steps to create Scan mode:

1. Create a new script in your **Project** Scripts/ folder by *right-clicking* and selecting **Create | C# Script** and name the script ScanMode.

2. Drag the ScanMode script onto the **Scan Mode** game object in the **Hierarchy** window.

3. *Double-click* the ScanMode script file to open it for editing and write it as follows:

```
using UnityEngine;
using UnityEngine.XR.ARFoundation;

public class ScanMode : MonoBehaviour
{
    [SerializeField] ARPlaneManager planeManager;

    void OnEnable()
    {
        UIController.ShowUI("Scan");
    }

    void Update()
    {
        if (planeManager.trackables.count > 0)
        {
            InteractionController.EnableMode("Main");
        }
    }
}
```

4. Drag the **AR Session Origin** object from the **Hierarchy** window onto the **Scan Mode | Plane Manager** slot.

When Scan mode is enabled, the **Scan UI** panel is shown. Then, it waits for at least one trackable plane has been detected by the AR system by checking for `planeManager.trackables.count > 0` before switching to Main mode.

The MainMode script

Main mode, as its name implies, is the main operating mode of the application. It may display the main menu, for example, and handle main user interactions. For our default framework, there's not much to do yet apart from display the Main UI panel.

Perform the following steps to create Main mode:

1. Create a new script in your Project's `Scripts/` folder by *right-clicking* and selecting **Create | C# Script** and name the script `MainMode`.

2. Drag the `MainMode` script onto the **Main Mode** game object in the **Hierarchy** window.

3. *Double-click* the `MainMode` script file to open it for editing and write it as follows:

```
using UnityEngine;

public class MainMode : MonoBehaviour
{
    void OnEnable()
    {
        UIController.ShowUI("Main");
    }
}
```

Lastly, we define NonAR mode.

The NonARMode script

NonAR mode will be enabled when the device you're running does not support AR. You might simply notify the user that the app cannot run, and gracefully exit. Alternatively, you may continue to run the app without AR capabilities if that makes sense for your project.

Perform the following steps to create a NonAR mode placeholder:

1. Create a new script in your Project's `Scripts/` folder by *right-clicking and selecting* **Create | C# Script,** and name the script `NonARMode`.

2. Drag the `NonARMode` script onto the **NonAR Mode** game object in the **Hierarchy** window.

3. *Double-click* the `NonARMode` script file to open it for editing and write it as follows:

```
using UnityEngine;

public class NonARMode: MonoBehaviour
{
    void OnEnable()
    {
        UIController.ShowUI("NonAR");
    }
}
```

That about does it. We've created a hierarchy with each of the interaction modes as children of **Interaction Controller**. To enable a mode, you'll call `InteractionController.EnableMode()`, which disables the current mode and activates a new one. When a mode is enabled, its mode script begins running, showing its UI, and potentially interacting with the user until specific conditions are met, and then transitions to a different mode. Let's try running the scene on your device.

Testing it out

Now is a good time to **Build And Run** the scene to make sure things are working as expected so far. Perform the following steps:

1. First, be sure to save your work by using **File | Save**.

2. Select **File | Build Settings** to open the **Build Settings** window.

3. Click **Add Open Scenes** to add the `ARFramework` scene to **Scenes In Build**, and ensure it is the only scene in the list with a checkmark.

4. Ensure that your target device is plugged into a USB port and that it is ready.

5. Click **Build And Run** to build the project.

Once the project builds without errors and launches on your device in **Startup** mode. You'll first see the words **Initializing…** from the **Startup** UI panel.

Once the AR Session is started, the app transitions to Scan mode and you will see the words **Scanning… Please move device slowly**.

Once a horizontal plane is being tracked, Scan mode transitions to Main mode. You will then see on the screen the words **Main Mode Running….**

If all goes well, the framework is working as intended. To accomplish this, we have implemented the Canvas UI and child panels for the user interface. We have implemented the Interaction Controller and child mode controllers with scripts that implement the UI and interactions required in each mode. And it's all wired together. This is a basic framework for an AR project that we will use for projects in this book.

There are many ways in which we can improve and build on this framework. For one, we can make the UI a little more interesting by replacing some of the text prompts with animated graphics from the AR Onboarding UX from Unity.

Using the Unity onboarding UX assets

Unity provides a set of AR onboarding UX assets useful for prompting users in an AR application. **Onboarding** refers to the user experience when your app starts up and prompts the user to interact with AR features. First, I'll explain some of what this package provides. Then we'll prepare the assets for use in our own projects.

Introducing the onboarding assets

The onboarding UX assets are part of the AR Foundation Demos project found at `https://github.com/Unity-Technologies/arfoundation-demos`. (This is different from the *AR Foundation Samples* project we explored in *Chapter 2, Your First AR Scene*). And its documentation can be found on that project's GitHub page.

The onboarding UX assets include icons and video graphics to prompt the user when scanning is required. It automatically tells the user the reasons why tracking may be failing, such as the room is too dark, or the camera view does not see sufficient details. It provides components to manage that process that are composed into an example prefab, named **ScreenspaceUI**, which can be customized to the look and feel of your own project.

For example, when the app is scanning, you can use an animated graphic prompt to *Move Device Slowly* while scanning the room. If there's a problem, it will display the reason, as shown in the left-side panel of the following image (where I have my finger covering the camera lens). It says **Look for more textures or details in the area**. If you want to prompt the user to tap the screen to place an object, there's a **Tap to Place** animated graphic, and so on:

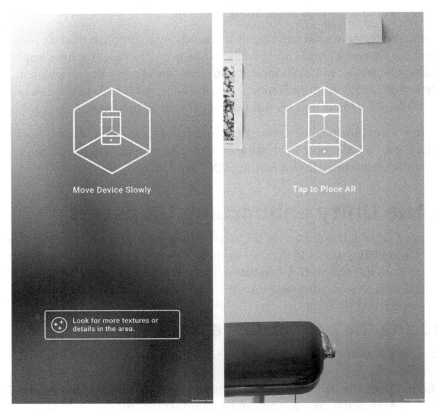

Figure 4.9 – Using the onboarding UX assets

Furthermore, the package supports localization of the text prompts, should your project require multi-language support for various countries. It also includes some good default assets for visualizing AR planes and point clouds that you can use.

The package includes the following components.

- **ARUX Animation Manager**: This displays instructional graphic animations to prompt the user to find a plane or tap to place, for example.

- **ARUX Reasons Manager**: This checks the AR Session's status and displays reasons why tracking may be failing as hints to the user.

- **Localization Manager**: This supports localized text and graphics for adapting the instructional and reasons UI to different languages.

- **UI Manager**: This is an example script for managing the user workflow.

> **Info: The UI Manager script is an example script**
>
> The `UIManager` script from the AR Foundation Demos project is a useful control script, but it is only an example of how to interface with the `ARUXAnimationManager`. Reading the script is informative but not reusable. In our framework, we have implemented our own solution for the user flow that replaces the `UIManager` script.

UI Manager lets you set up one or two goals via the **Inspector** window. A goal may be **Found a Plane** or **Placed an Object**. You then set the instructional UI to prompt the user to perform the current activity until its goal has been completed.

Preparing the Unity AR onboarding assets

While the onboarding UX assets are also available as a package in the Unity Asset Store, I recommend you clone the GitHub project version because it has more examples and assets, including **Universal Render Pipeline** (**URP**) shader-graph shaders. Both versions are full Unity projects, so either way, you will need to open it in a new Unity project and then export the assets into a package that you can import into your own projects.

We will clone the project and then export the AR Foundation Demos assets into a `.unitypackage` file that we can import into our own project. I will also provide a copy of this Unity package with the files for this book in the GitHub repository.

To clone the project and export the folders we want, perform the following steps:

1. Clone a copy of the project from GitHub to your local machine. The project can be found at `https://github.com/Unity-Technologies/arfoundation-demos`. Please use whatever cloning method you prefer, for example, GitHub Desktop (`https://desktop.github.com/`) or Command Line (`https://git-scm.com/download/`).

2. Open the **Unity Hub** application on your desktop.

3. Add the project to **Unity Hub** by selecting **Projects | Add**, navigating to the cloned project's root folder, and then press **Select Folder**.

4. In the **Unity Hub** projects list, if you see a yellow warning icon indicating that the Unity version used by the cloned project is not presently installed on your system, use the **Unity Version** selection to choose a newer version of the editor that you do have installed (preferably the same major release number).

5. Open the project by selecting it from the **Unity Hub** projects list.

6. We're going to move selected folders into a root folder named `ARFoundationDemos` that we can export into a package.

    In Unity, in the **Project** window, create a new folder using the + button in the top-left of the **Project** window and name it `ARFoundationDemos`.

7. With your mouse, move the following four folders into this `ARFoundationDemos/` folder: **AddressableAssetsData**, **Common**, **Shaders**, and **UX**.

8. In the **Project** window, *right-click* on the `ARFoundationDemos/` folder and select **Export Package**.

9. The **Exporting Package** window will open. Click **Export**.

10. Choose a directory outside of this project's root and name the file (such as `ARF-OnboardingUX`). Then, click **Save**.

Before you close the `ARFoundationDemos` project, you may want to look in the **Package Manager** window and note the **AR Foundation** package version used in the given project, to make sure your own project uses the same or later version of AR Foundation.

You can close the `ARFoundationDemos` project now. You now have an asset package you can use in this and other projects.

Installing dependency packages

The AR onboarding UX has some dependencies on other Unity packages that you must install in your own project: *Addressables* and *Localization*. Open your AR project and install them now.

The Addressable Asset system simplifies loading assets at runtime with a unified scheme. Assets can be loaded from any location with a unique address, whether they reside in your application or on a content delivery network. Assets can be accessed via direct references, traditional asset bundles, or `Resource` folders. The **Addressables** package is required by the onboarding UX assets. To learn more, see `https://docs.unity3d.com/Packages/com.unity.addressables@1.16/manual/index.html`.

To import the **Addressables** package, perform the following steps:

1. Open the **Package Manager** window by using **Window | Package Manager**.

2. Ensure **Unity Registry** is selected from the **Packages** filter dropdown in the upper-left corner of the **Package Manager** window.

3. Search for `Addressables` using the search text input field in the upper-right corner of the **Package Manager** window.

4. Select the **Addressables** package and click **Install**.

The **Addressables** package is now installed.

The **Localization** package translates text strings and other assets into local languages. See `https://docs.unity3d.com/Packages/com.unity.localization@1.0/manual/index.html`. To import the **Localization** package, perform the following steps (these steps may have changed by the time you read this):

1. If you have not already done so, enable **Preview Packages** by navigating to the **Edit | Project Settings | Package Manager** settings and checking the **Enable Preview Packages** checkbox.

2. Then, in the **Package Manager** window, use the + button in the top-left corner and select **Add Package From Git URL**.

3. Then, type `com.unity.localization` to begin installing the package.

> **Info: Using Preview packages and Git URLs**
>
> As I write this, the **Localization** package is in *preview*, that is, not yet fully released by Unity. Also, it is not yet included in the Unity package registry. To enable preview packages, you must click **Enable Preview Packages** in **Project Settings**. Also if a package is not included in the built-in Unity registry, you can add a package from a Git URL, from disk, or from a tarball file.

The **Localization** package is now installed. We can now install the AR onboarding UX assets themselves.

Importing the OnboardingUX package

We saved the assets exported from the AR Foundation Demos project into a file named `OnboardingUX.unitypackage`. Importing the package is straightforward. Follow these steps to add it to your project. Back in your own Unity project, do the following:

1. Select **Assets | Import Package | Custom Package**. Alternatively, drag the `OnboardingUX.unitypackage` file from your Explorer or Finder directly into the Unity **Project** window.

2. In the **Import Unity Package** window, click **Import**.

3. The assets include materials that use the built-in render pipeline. Since our project is using the URP, you need to convert the materials by selecting **Edit | Render Pipeline | Universal Render Pipeline | Upgrade Project Materials to URP**.

> Tip: SphereObject shadow material in the URP
>
> The **SphereObject** prefab that comes with the onboarding UX demo assets is configured to cast a shadow using the built-in render pipeline, not the URP. As such, the shadow appears as a missing shader and is colored magenta. To fix this, locate the material named **ShadowMat** in the `ARFoundationDemos/ Common/Materials/` folder and, in the **Inspector** window, change its **Shader**, using the drop-down menu, to **ShaderGraphs/BlurredShadowPlane**.

The onboarding UX assets are now imported into your project. We can now add it to our framework scene.

Currently, our app renders a UI panel with text to prompt the user to scan the environment. This panel is a game object that is enabled when needed. Basically, we want to replace the panel text with animated graphics.

Writing the AnimatedPrompt script

Let's start by writing a new script, `AnimatedPrompt`, that displays a specific animation when it is enabled and hides the animation when disabled:

1. Create a new script in your **Project** `Scripts/` folder by *right-clicking and selecting* **Create | C# Script** and name the script `AnimatedPrompt`.

2. *Double-click* the file to open it for editing and replace the default content, starting with the following declarations:

```
using UnityEngine;

public class AnimatedPrompt : MonoBehaviour
{
    public enum InstructionUI
    {
        CrossPlatformFindAPlane,
        FindAFace,
        FindABody,
        FindAnImage,
        FindAnObject,
        ARKitCoachingOverlay,
        TapToPlace,
        None
    };

    [SerializeField] InstructionUI instruction;
    [SerializeField] ARUXAnimationManager
        animationManager;

    bool isStarted;
```

In this script, we declare a public property, `instruction`, whose value is an `enum` `InstructionUI` type that indicates which animation to play (borrowed from the `UIManager` script from the onboarding assets, to be consistent).

3. When the script is started or enabled, it will initiate the animated graphics. Inversely, when the object is disabled, the graphics are turned off:

```
    void Start()
    {
        ShowInstructions();
        isStarted = true;
    }

    void OnEnable()
```

```
    {
        if (isStarted)
            ShowInstructions();
    }


    void OnDisable()
    {
        animationManager.FadeOffCurrentUI();
    }
```

I've added a fix to ensure the animation does not restart when both Start and OnEnable are called at the start.

4. When the script is enabled, it calls the helper function, ShowInstructions, which calls a corresponding function in ARUXAnimationManager:

```
    void ShowInstructions()
    {
        switch (instruction)
        {
            case InstructionUI.CrossPlatformFindAPlane:
                animationManager.
                    ShowCrossPlatformFindAPlane();
                break;
            case InstructionUI.FindAFace:
                animationManager.ShowFindFace();
                break;
            case InstructionUI.FindABody:
                animationManager.ShowFindBody();
                break;
            case InstructionUI.FindAnImage:
                animationManager.ShowFindImage();
                break;
            case InstructionUI.FindAnObject:
                animationManager.ShowFindObject();
                break;
            case InstructionUI.TapToPlace:
                animationManager.ShowTapToPlace();
```

```
            break;
        default:
            Debug.LogError("instruction switch
            missing, please edit AnimatedPrompt.cs "
            + instruction);
            break;
    }
  }
}}
```

Now we can add this to our scene.

Integrating the onboarding graphics

To integrate the onboarding graphics, we can add the demo prefab (unfortunately named ScreenspaceUI) from the AR Foundation Demos package. Follow these steps:

1. In the **Project** window, navigate to the ARFoundationDemos/UX/Prefabs/ folder and drag the **ScreenspaceUI** prefab into the **Hierarchy** window root of the scene.

2. Give it a more indicative name; rename the object OnboardingUX.

3. Our framework replaces the demo **UI Manager** component, so you should remove this.

   With the **OnboardingUX** object selected in **Hierarchy**, click the *3-dot context menu* in the top-right corner of the **UI Manager** component in the **Inspector** window and select **Remove Component**.

We can now use AnimatedPrompt to replace the text in our UI prompt panels. To use it, perform the following steps:

1. In the **Hierarchy** window, select the **Scan UI** panel object, *right-click* **Create Empty**, and rename the new object Animated Prompt.

2. With the **Animated Prompt** object selected, drag the new AnimatedPrompt script from the **Project** window onto the object.

3. Set the **Animated Prompt | Instruction** to **Cross-Platform Find A Plane**.

4. From the **Hierarchy** window, drag the **OnboardingUX** object into the **Inspector** window and drop it on to the **Animation Manager** slot.

5. You can disable the **Text (TMP)** child element of **Scan Prompt Panel** so that it won't be rendered.

If you **Build And Run** the project again, when it enters Scan mode, you will be greeted with nice, animated graphics instead of the text prompt.

With a working AR user framework, let's make this scene into a template that we can use when creating new scenes.

Creating a scene template for new scenes

We can save this **ARFramework** scene we've been working on as a template to use for starting new scenes in this Unity project. To create a scene template, perform the following steps.

1. With the **ARFramework** scene open, select **File | Save As Scene Template**.

2. In the **Save** window, navigate to your `Scenes/` folder, verify the template name (`ARFramework.scenetemplate`), and then press **Save**.

3. Subsequently, when you want to start a new AR scene, use this template. By default, Unity will duplicate any dependencies within the scene into a separate folder. In our case, this is generally *not* what we want to do.

   To prevent cloning the scene dependencies when the template is used, click on this new scene template file in your **Project** `Assets/` window.

4. In its **Inspector** window, in the **Dependencies** panel, uncheck each of the assets you do not want to be cloned and want to be shared between your scenes. In our case, we do not want to clone any, so use the checkbox at the top of the **Clone** column to change the checkboxes in bulk.

> **Tip: Updating a template scene**
>
> A Unity scene template contains metadata used when selecting and instantiating a new scene via the **File | New Scene** menu. It does not include the scene's GameObjects. Rather, it contains a reference to your prototype Unity scene. If you want to modify this prototype, don't re-save the scene as a new template. You simply edit the scene it is referencing. In this case, that will be the scene named `ARFramework`. Just remember to check the **Dependencies** list in the template if you've added any new assets to the scene as these will default to be cloned.

To use the template when creating a new scene in this project, use **File | New Scene** as usual. The dialog box will now contain the **ARFramework** template as an option. Select the location in your assets folder and press **Create**. If the template specifies any assets to be cloned, those copies will be added to a subfolder with the same name as the new scene.

We are now ready to build upon the work we did in this chapter, using the **ARFramework** template for new project scenes.

Summary

In this chapter, we developed a framework for building AR applications and saved it as a template we can use for projects in this book. The framework provides a state-machine structure for implementing modes and identifying the conditions when to transition to a different mode. The framework also offers a controller-view design pattern where, when a mode is active, its corresponding UI is visible, keeping the mode control objects separate from the UI view objects.

For the framework template, we implemented four modes: Startup mode, Scan mode, Main mode, and NonAR mode, along with four UI panels: Startup UI, Scan UI, Main UI, and NonAR UI. Scan mode uses the onboarding UX assets from the AR Foundation Demos project to prompt the user to scan for trackable features and report problems with detection and the AR session.

In the next chapter, I will demonstrate the use of this framework with a simple demo project and then build upon the framework more extensively in subsequent chapters.

5
Using the AR User Framework

In this chapter, we will learn how to use the **Augmented Reality** (**AR**) user framework that we set up in the previous chapter, *Chapter 4, Creating an AR User Framework*. Starting with the ARFramework scene template, we will add a main menu for placing virtual objects in the environment. If you skipped that chapter or just read through it, you can find the scene template and assets in the files provided on this book's GitHub repository.

For this project, we'll extend the framework with a new *PlaceObject-mode* that prompts the user to tap to place a virtual object in the room. The user will have a choice of objects from the main menu.

In the latter half of the chapter, I'll discuss some advanced AR application issues including making an AR-optional project, determining whether a device supports a specific AR feature, and adding localization to your **User Interface** (**UI**).

This chapter will cover the following topics:

- Planning the project
- Starting with the ARFramework scene template
- Adding a main menu
- Adding PlaceObject mode and instructional UI

- Wiring the menu buttons

- Doing a Build And Run

- Hiding tracked objects when not needed

- Making an AR-optional project

- Determining whether a device supports specific AR features at runtime

- Adding localization features to a project

By the end of the chapter, you'll be more familiar with the AR user framework developed for this book, which we'll use in subsequent chapters as we build a variety of different AR application projects.

Technical requirements

To implement the project in this chapter, you need Unity installed on your development computer, with a mobile device connected that supports AR applications (see *Chapter 1, Setting Up for AR Development,* for instructions), including the following:

- Universal Render Pipeline

- Input System package

- XR Plugin for your target device

- AR Foundation package

We assume you have installed the assets from the Unity **arfoundation-samples** project imported from `ARF-samples.unitypackage` created in *Chapter 2, Your First AR Scene.*

Also from *Chapter 2, Your First AR Scene,* we created an **AR Input Actions** asset that we'll use in this project, containing an **action map** named **ARTouchActions** including (at least) a **PlaceObject** action.

We also assume you have the `ARFramework` scene template created in *Chapter 4, Creating an AR User Framework,* along with all the prerequisite Unity packages detailed at the beginning of *Chapter 4 Creating an AR User Framework.* A copy of the template and assets can be found in this book's GitHub repository at `https://github.com/PacktPublishing/Augmented-Reality-with-Unity-AR-Foundation` (not including the third-party packages that you should install yourself).

The AR user framework requires the following prerequisites, as detailed in *Chapter 4, Creating an AR User Framework,* including the following:

- The Addressables package

- The Localization package

- TextMesh Pro

- The DOTween package from the Asset Store

- The Serialized Dictionary Lite package from the Asset Store

The completed scene for this chapter can also be found in the GitHub repository.

Planning the project

For this project, we'll create a simple demo AR scene starting with the `ARFramework` scene template and building up the user framework structure we have set up.

With the framework, when the app first starts, Startup-mode is enabled and the AR Session is initialized. Once the session is ready, it transitions to Scan-mode.

If the AR Session determines that the current device does not support AR, Scan-mode will transition to NonAR-mode instead. Presently this just puts a text message on the screen. See the *Making an AR-optional project* section near the end of this chapter for more information.

In Scan-mode, the user is prompted to use their device camera to slowly scan the room until AR features are detected, namely, horizontal planes. The **ScanMode** script checks for any tracked planes and then transitions to Main-mode.

Given this, our plan is to add the following features:

- The AR session will be configured to detect and track horizontal planes. We'll also render point clouds.

- Main-mode will show a main menu with buttons that lets the user choose objects to place in the real-world environment. You can find your own models to use here, but we'll include three buttons for a cube, a sphere, and a virus (created in *Chapter 2, Your First AR Scene*).

- When a place-object button is selected, it will enable a new PlaceObject-mode that prompts the user to tap to place the objects onto a detected plane.

- Tapping on a tracked horizontal plane will create an instance of the object in the scene. The app then goes back to Main-mode.

- Tracked AR features (planes and point clouds) will be hidden in Main-mode, and visible in PlaceObject-mode.

I have chosen to provide a cube, a sphere, and a virus (the virus model was created in *Chapter 2, Your First AR Scene*). Feel free to find and use your own models instead. The prefab assets I will be using are the following:

- **AR Placed Cube** (found in the `Assets/ARF-samples/Prefabs/` folder)
- **AR Placed Sphere** (found in the `Assets/ARF-samples/Prefabs/` folder)
- **Virus** (found in `Assets/_ARFBookAssets/Chapter02/Prefabs/` folder)

This is a simple AR demo that will help you become more familiar with the AR user framework we developed and will use in subsequent projects in this book.

Let's get started.

Starting with the ARFramework scene template

To begin, we'll create a new scene named `FrameworkDemo` using the `ARFramework` scene template, using the following steps:

1. Select **File | New Scene**.
2. In the **New Scene** dialog box, select the **ARFramework** template.
3. Press **Create**.
4. Select **File | Save As**. Navigate to the `Scenes/` folder in your project's `Assets` folder, give it the name `FrameworkDemo`, and press **Save**.

> **Note: Unintended clone dependencies**
>
> When creating a new scene from a scene template, if you're prompted right away for a name to save the file under, this indicates your scene template has some clone dependencies defined. If this is not your intention, cancel the creation, select the template asset in your Project window, and ensure all the **Clone** checkboxes are cleared in the **Dependencies** list. Then try creating your new scene again.

The new AR scene already has the following game objects included from the template:

- The **AR Session** game object
- The **AR Session Origin** rig with the raycast manager and plane manager components.

- **UI Canvas** is a screen space canvas with Startup UI, Scan UI, Main UI, and NonAR UI child panels. It also has the UI Controller component script that we wrote.

- **Interaction Controller** is a game object with the Interaction Controller component script we wrote that helps the app switch between interaction modes, including Startup, Scan, Main, and NonAR modes. It also has a **Player Input** component configured with the **AR Input Actions** asset we previously created.

- The **OnboardingUX** prefab from the AR Foundation Demos project that provides AR session status and feature detection status messages, and animated onboarding graphics prompts.

Set up the app title now as follows:

1. In the **Hierarchy** window, unfold the **UI Canvas** object, and unfold its child **App Title Panel**.
2. Select the **Title Text** object.
3. In its **Inspector**, change its text content to `Place Object Demo`.

The default AR Session Origin already has an AR Plane Manager component. Let's ensure it's only detecting horizontal planes. Let's add a point cloud visualization too. Follow these steps:

1. In the **Hierarchy** window, select the **AR Session Origin** object.
2. In the **Inspector**, set the **AR Plane Manager | Detection Mode** to **Horizontal** by first selecting **Nothing** (to clear the list) and then selecting **Horizontal**.
3. Click the **Add Component** button, search for `ar point cloud`, then add an **AR Point Cloud Manager** component.
4. Find a point cloud visualizer prefab and set the **Point Cloud Prefab** slot (for example, **AR Point Cloud Debug Visualizer** can be found in the `Assets/ARF-samples/Prefabs/` folder).
5. Save your work with **File | Save**.

We've created a new scene based on the ARFramework template and added AR trackables managers for point clouds and horizontal planes. Next, we'll add the main menu.

Adding a main menu

The main menu UI resides under the Menu UI panel (under UI Canvas) in the scene hierarchy. We will add a menu panel with three buttons to let you add a cube, a sphere, and a virus. We'll create a menu sub-panel and arrange the menu buttons horizontally. Follow these steps:

1. In the **Hierarchy**, unfold the **UI Canvas**, and unfold its child **Main UI** object.

2. First, remove the temporary **Main** mode text element. *Right-click* the child **Text** object and select **Delete**.

3. *Right-click* the **Menu UI** and select **UI | Panel**, then rename it `Main Menu`.

4. On the **Main Menu** panel, use the **Anchor Presets** to set **Bottom-Stretch**, and use *Shift* + *Alt* + click **Bottom-Stretch** to make a bottom panel. Then set **Rect Transform | Height** to `175`.

5. I set my background **Image | Color** to opaque white with **Alpha:** `255`.

6. Select **Add Component**, search `layout`, then select **| Horizontal Layout Group**.

7. On the **Horizontal Layout Group** component check the **Control Child Size | Width** and **Height** checkboxes (leave the others at their default values, **Use Child Scale** unchecked, and **Child Force Expand** checked). The Main Menu panel looks like this in the Inspector:

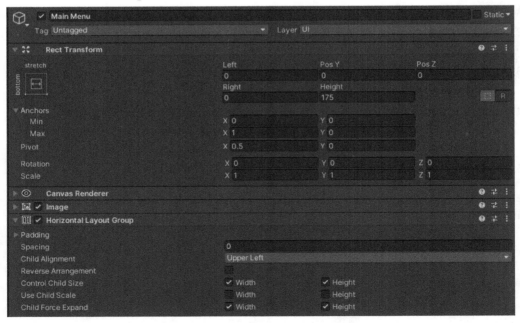

Figure 5.1 – The Main Menu panel settings

Now we'll add three buttons to the menu using the following steps:

1. *Right-click* the **Main Menu**, select **UI | Button – TextMeshPro**, and rename it to `Cube Button`.

2. Select its child text object, and set the **Text** value to `Cube` and **Font Size** to `48`.

3. *Right-click* the **Cube Button** and select **Duplicate** (or press *Ctrl + D*). Rename it `Sphere Button` and change its text to `Sphere`.

4. Repeat *step 3* again, renaming it `Virus Button`, and changing the text to `Virus`.

The resulting scene hierarchy of the Main Menu is shown in the following screenshot:

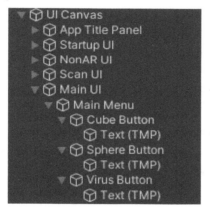

Figure 5.2 – Main Menu hierarchy

I decided to go further and add a sprite image of each model to the buttons. I created the images by screen-capturing a view of each model, edited them with Photoshop, saved them as PNG files, and in Unity made sure the image's **Texture Type** is set to **Sprite (2D and UI)**. I then added a child **Image** element to the buttons. The result is as shown in the following image of my menu:

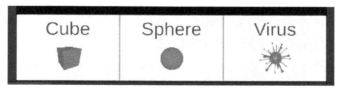

Figure 5.3 – Main Menu with icon buttons

Thus far we have created a Main Menu panel with menu buttons under the Main UI. When the app goes into Main-mode, this menu will be displayed.

Next, we'll add a UI panel that prompts the user to tap the screen to place an object into the scene.

Adding PlaceObject-mode with instructional UI

When the user picks an object from the main menu, the app will enable PlaceObject-mode. For this mode, we need a UI panel to prompt the user to tap the screen to place the object. Let's create the UI panel first.

Creating the PlaceObject UI panel

The **PlaceObject UI** panel should be similar to the **Scan UI** one, so we can duplicate and modify it using the following steps:

1. In the **Hierarchy** window, unfold the **UI Canvas**.

2. *Right-click* the **Scan UI** game object and select **Duplicate**. Rename the new object PlaceObject UI.

3. Unfold **PlaceObject UI** and select its child **Animated Prompt**.

4. In the **Inspector**, set the **Animated Prompt | Instruction** to **Tap To Place**. The resulting component is shown in the following screenshot:

Figure 5.4 – Animated Prompt settings for the PlaceObject UI panel

5. Now we add the panel to the UI Controller.

    In the **Hierarchy**, select the **UI Canvas** object.

6. In the **Inspector**, at the bottom-right of the **UI Controller** component, click the + button to add an item to the UI Panels dictionary.

7. Enter PlaceObject as text in the **Id** field.

8. Drag the **PlaceObject UI** game object from the **Hierarchy** onto the **Value** slot. The UI Controller component now looks like the following:

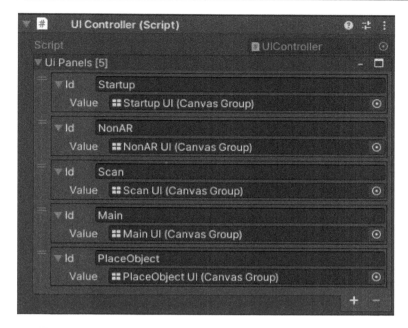

Figure 5.5 – UI Controller's UI Panels list with PlaceObject added

We added an instructional user prompt for the **PlaceObject** UI. When the user chooses to add an object to the scene, this panel will be displayed. Next, we'll add the **PlaceObject** mode and script.

Creating the PlaceObject mode

To add a mode to the framework, we create a child GameObject under the **Interaction Controller** and write a mode script. The mode script will show the mode's UI, handle any user interactions, and then transition to another mode when it is done. For PlaceObject-mode, it will display the **PlaceObject UI** panel, wait for the user to tap the screen, instantiate the prefab object, and then return to Main-mode.

Let's write the `PlaceObjectMode` script as follows:

1. Begin by creating a new script in your **Project** `Scripts/` folder using *right-click | ***Create C# Script**, and name the script `PlaceObjectMode`.

2. *Double-click* the file to open it for editing and replace the default content, starting with the following declarations:

```
using System.Collections;
using System.Collections.Generic;
using UnityEngine;
```

```
using UnityEngine.InputSystem;
using UnityEngine.XR.ARFoundation;
using UnityEngine.XR.ARSubsystems;

public class PlaceObjectMode : MonoBehaviour
{
    [SerializeField] ARRaycastManager raycaster;
    GameObject placedPrefab;
    List<ARRaycastHit> hits = new List<ARRaycastHit>();
```

The script will use APIs from ARFoundation and ARSubsystems so we specify these in the using statements at the top of the script. It will use the ARRaycastManager to determine which tracked plane the user has tapped. Then it will instantiate the placedPrefab into the scene.

3. When the mode is enabled, we will show the **PlaceObject UI** panel:

```
void OnEnable()
{
    UIController.ShowUI("PlaceObject");
}
```

4. When the user selects an object from the Main Menu, we need to tell PlaceObjectMode which prefab to instantiate, given the following code:

```
public void SetPlacedPrefab(GameObject prefab)
{
    placedPrefab = prefab;
}
```

5. Then when the user taps the screen, the Input System triggers an OnPlaceObject event (given the **AR Input Actions** asset we previously set up), using the following code:

```
public void OnPlaceObject(InputValue value)
{
    Vector2 touchPosition = value.Get<Vector2>();
    PlaceObject(touchPosition);
}

void PlaceObject(Vector2 touchPosition)
```

```
        {
            if (raycaster.Raycast(touchPosition, hits,
                TrackableType.PlaneWithinPolygon))
            {
                Pose hitPose = hits[0].pose;
                Instantiate(placedPrefab, hitPose.position,
                    hitPose.rotation);

                InteractionController.EnableMode("Main");
            }
        }
    }
```

When a touch event occurs, we pass the touchPosition to the PlaceObject function, which does a Raycast to find the tracked horizontal plane. If found, we Instantiate the placedPrefab at the hitPose location and orientation. And then the app returns to Main-mode.

6. Save the script and return to Unity.

We can now add the mode to the Interaction Controller as follows:

1. In the **Hierarchy** window, right-click the **Interaction Controller** game object and select **Create Empty**. Rename the new object PlaceObject Mode.

2. Drag the **PlaceObjectMode** script from the **Project** window onto the **PlaceObject Mode** object adding it as a component.

3. Drag the **AR Session Origin** object from the **Hierarchy** onto the **Place Object Mode | Raycaster** slot.

    Now we'll add the mode to the **Interaction Controller**.

4. In the **Hierarchy**, select the **Interaction Controller** object.

5. In the **Inspector**, at the bottom-right of the **Interaction Controller** component, click the + button to add an item to the **Interaction Modes** dictionary.

6. Enter the PlaceObject text in the **Id** field.

7. Drag the **PlaceObject Mode** game object from the **Hierarchy** onto the **Value** slot. The **Interaction Controller** component now looks like the following:

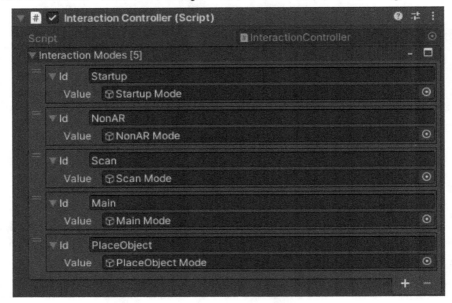

Figure 5.6 – The Interaction Controller's Interaction Modes list with PlaceObject added

We have now added a **PlaceObject Mode** to our framework. It will be enabled by the Interaction Controller when **EnableMode("PlaceObject")** is called by another script or, in our case, by a main menu button. When enabled, the script shows the **PlaceObject** instructional UI, then listens for an `OnPlaceObject` input action event. Upon the input event, we use Raycast to determine where in the 3D space the user wants to place the object, then the script instantiates the prefab and returns to Main-mode.

The final step is to wire up the main menu buttons.

Wiring the menu buttons

When the user presses a main menu button to add an object to the scene, the button will tell `PlaceObjectMode` which prefab is to be instantiated. Then PlaceObject mode is enabled, which prompts the user to tap to place the object and handles the user input action. Let's set up the menu buttons now using the following steps:

1. Unfold the **Main Menu** game object in the **Hierarchy** by navigating to **UI Canvas / Main UI / Main Menu** and select the **Cube Button** object.

2. In its **Inspector**, on the **Button** component, in its **OnClick** section, press the + button in the bottom right to add an event action.

3. From the **Hierarchy**, drag the **PlaceObject Mode** object onto the **OnClick Object** slot.

4. In the **Function** selection list, choose **PlaceObject Mode | SetPlacedPrefab**.

5. In the **Project** window, locate a cube model prefab to use. For example, navigate to your `Assets/ARF-samples/Prefabs/` folder and drag the **AR Placed Cube** prefab into the **Game Object** slot for this click event in **Inspector**.

6. Now let the button enable PlaceObject Mode. In its **Inspector**, on the **Button** component, in its **OnClick** section, press the + button in the bottom right to add another event action.

7. From the **Hierarchy**, drag the **Interaction Controller** object onto the **OnClick** event's **Object** slot.

8. In the **Function** selection list, choose **InteractionController | EnableMode**.

9. In the string parameter field, enter `PlaceObject`.

The Cube Button object's Button component now has the following **OnClick** event settings:

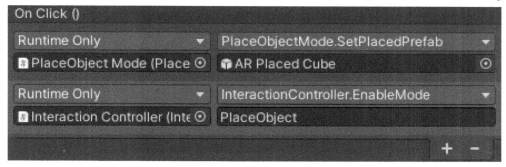

Figure 5.7 – The OnClick events for the Cube Button

Repeat these steps for the Sphere Button and Virus Button. As a shortcut we can copy/paste the component settings as follows:

1. With the **Cube Button** selected in the **Hierarchy**, over in the **Inspector**, click the three-dot context menu for the **Button** component, and select **Copy Component**.

2. In the **Hierarchy**, select the **Sphere Button** object.

3. In its **Inspector**, click the three-dot context menu for the **Button** component, and select **Paste Component Values**.

4. In the **Project** window, locate a sphere model prefab to use. For example, navigate to your `Assets/ARF-samples/Prefabs/` folder and drag the **AR Placed Sphere** prefab into the **Game Object** slot for this click event in **Inspector**.

5. Likewise, repeat *steps 1-4* for the **Virus Button**, and set the GameObject to the **Virus** prefab (perhaps located in your own `Prefabs` folder).

6. Save your work using **File | Save**.

Everything should be set up now. We created a new scene using the **ARFramework** template, added a main menu with buttons, added the PlaceObject-mode with instructional user prompt, wrote the `PlaceObjectMode` script that handles user input actions and instantiates the prefab, and wired it all up to the main menu buttons. Let's try it out!

Performing a Building and Run

To build and run the project, use the following steps:

1. Open the **Build Settings** window using **File | Build Settings**.

2. Click the **Add Open Scenes** button if the current scene (`FrameworkDemo`) is not already in the **Scenes In Build** list.

3. Ensure that the `FrameworkDemo` scene is the only one checked in the **Scenes In Build** list.

4. Click **Build And Run** to build the project.

When the project builds successfully, it starts up in Startup-mode while the AR Session is initializing. Then it goes into Scan-mode that prompts the user to scan the environment, until at least one horizontal plane is detected and tracked. Then it goes into Main-mode and displays the main menu. Screen captures of the app running on my phone in each of these modes are shown in the following figure:

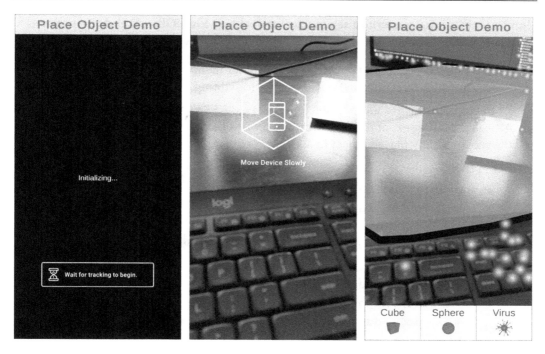

Figure 5.8 – Screen captures of Startup-mode, Scan-mode, and Main-mode

On pressing one of the menu buttons, the app goes into PlaceObject-mode, prompting the user to tap to place an object. Tapping the screen instantiates the object at the specified location in the environment. Then the app returns to Main-mode.

We now have a working demo AR application for placing various virtual objects onto horizontal surfaces in your environment. One improvement might be to hide the trackable objects in Main-mode and only display them when needed in PlaceObject-mode.

Hiding tracked objects when not needed

When the app first starts tracking, we show the trackable planes and point clouds. This is useful feedback to the user when the app first starts and subsequently when placing an object. But once we have objects placed in the scene, these trackable visualizations can be distracting and unwanted. Let's only show the object while in PlaceObject-mode and hide them after at least one virtual object has been placed.

In AR Foundation, hiding the trackables requires two separate things: hiding the existing trackables that have already been detected, and preventing new trackables from being detected and visualized. We will implement both.

To implement this, we can write a separate component on **PlaceObject mode** that shows the trackables when enabled and hides them when disabled. Follow these steps:

1. Create a new C# script in your `Scripts/` folder named `ShowTrackablesOnEnable` and open it for editing.

2. At the top of the class, add variable references to `ARSessionOrigin`, `ARPlaneManager`, and `ARPointCloudManager`. Also, we will now remember the most recently placed object in `lastObject`, and initialize them in `Awake`, as follows:

```csharp
using UnityEngine;
using UnityEngine.XR.ARFoundation;


public class ShowTrackablesOnEnable : MonoBehaviour
{
    [SerializeField] ARSessionOrigin sessionOrigin;
    ARPlaneManager planeManager;
    ARPointCloudManager cloudManager;
    bool isStarted;

    void Awake()
    {
        planeManager =
            sessionOrigin.GetComponent<ARPlaneManager>();
        cloudManager = sessionOrigin.GetComponent
            <ARPointCloudManager>();
    }
    private void Start()
    {
        isStarted = true;
    }
```

I've also added an `isStarted` flag that we'll use to prevent the visualizers from being hidden when the app starts up.

> **Info: OnEnable and OnDisable can be called before Start**
>
> In the life cycle of a `MonoBehaviour` component, `OnEnable` is called when the object becomes enabled and active. `OnDisable` is called when the script object becomes inactive. `Start` is called on the first frame the script is enabled, just before `Update`. See `https://docs.unity3d.com/ScriptReference/MonoBehaviour.Awake.html`.
>
> In our app, it is possible for `OnDisable` to get called before `Start` (when we're initializing the scene from `InteractionController`). To prevent `ShowTrackables(false)` from getting called before the scene has started, we use an `isStarted` flag in this script.

3. We will show the trackables when the mode is enabled and hide them when disabled using the following code:

```
void OnEnable()
{
    ShowTrackables(true);
}

void OnDisable()
{
    if (isStarted)
    {
        ShowTrackables(false);
    }
}
```

4. These call `ShowTrackables`, which we implement as follows:

```
void ShowTrackables(bool show)
{
    if (cloudManager)
    {
        cloudManager.SetTrackablesActive(show);
        cloudManager.enabled = show;
    }
    if (planeManager)
    {
        planeManager.SetTrackablesActive(show);
```

```
                    planeManager.enabled = show;
        }
    }
}
```

Setting `SetTrackablesActive(false)` will hide all the existing trackables. Disabling the trackable manager component itself will prevent new trackables from being added. We check for null managers in case the component is not present in `ARSessionOrigin`.

5. Save the script.

6. Back in Unity, select the **PlaceObject Mode** game object in the **Hierarchy**.

7. Drag the `ShowTrackablesOnEnable` script onto the **PlaceObject Mode** object.

8. Drag the **AR Session Origin** object from the **Hierarchy** into the **Inspector** and drop it onto the **Show Trackables On Enable | Session Origin** slot.

9. Save the scene using **File | Save**.

Now when you click **Build And Run** again, the trackables will be shown when PlaceObject Mode is enabled, and will be hidden when disabled. Thus, the trackables will be visible when Main mode is first enabled but after an object has been placed and the app goes back to Main-mode, the trackables will be hidden. This is the behavior we desire. The PlaceObject-mode and subsequent Main-mode are shown in the following screen captures of the project running on my phone:

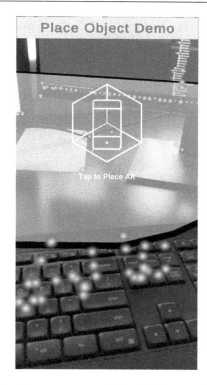

Figure 5.9 – Screen captures of PlaceObject-mode, and subsequent Main-mode with trackables hidden

---

Tip: Disable trackables by modifying the plane detection mode

To disable plane detection, the method I'm using is to disable the
manager component. This is the technique given in the example
`PlaneDetectionController.cs` script in the AR Foundation
Samples project. Alternatively, the Unity ARCore XR Plugin docs (
`https://docs.unity3d.com/Packages/com.unity.`
`xr.arcore@4.1/manual/index.html`) recommend disabling plane
detection by setting the `ARPlaneManager` detection mode to the value
`PlaneDetectionMode.None`.

We've now completed a simple AR project to place various virtual objects on horizontal planes detected in the environment, using our AR user framework.

Further improvements you could add to the project include the following:

- A reset button in the main menu to remove any virtual objects already placed in the scene.

- Only allow one instance of a virtual object to be placed in the scene at a time.

- The ability to move and resize an existing object (see *Chapter 7, Gallery: Editing Virtual Objects*).
- Can you think of more improvements? Let us know.

In the rest of this chapter, we'll discuss some advanced onboarding and user experience features you may want to include in your projects at a later time.

Advanced onboarding issues

In this section, we'll review some other issues related to AR onboarding, AR sessions, and devices, including the following:

- Making an AR-optional project
- Determining whether the device supports a specific AR feature
- Adding localization to your project

Making an AR-optional project

Some applications are intended to be run specifically using AR features and should just quit (after a friendly notification to the user) if it's not supported. But other applications may want to behave like an ordinary mobile app with an extra optional capability of supporting AR features.

For example, a game I recently created, Epoch Resources (available for Android at `https://play.google.com/store/apps/details?id=com.parkerhill.EpochResources&hl=en_US&gl=US`, and iOS at `https://apps.apple.com/us/app/epoch-resources/id1455848902`) is a planetary evolution incremental game with a 3D planet you mine for resources. It offers an optional AR-viewing mode where you can "pop" the planet into your living room and continue playing the game in AR, as shown in the following image.

Figure 5.10 – Epoch Resources is an AR-optional game

For an AR-optional application, your app will probably start up as an ordinary non-AR app. Then at some point the user may choose to turn on AR-specific features. That's when you'll activate the AR Session and handle the onboarding UX.

None of the projects in this book implement AR-optional so this is an informational discussion only. To start, you'll tell the XR Plugin that AR is optional by going to **Edit | Project Settings | XR Plug-in Management** and selecting **Requirement | Optional** (instead of **Required**) for each of your platforms (ARCore and ARKit are set separately).

You will need a mechanism for running with or without AR. One approach is to have separate AR and non-AR scenes that are loaded as needed (see `https://docs.unity3d.com/ScriptReference/SceneManagement.SceneManager.html`).

In the case of the Epoch Resources game, we did not create two separate scenes. Rather the scene contains two cameras, the normal default camera for non-AR mode and the AR Session Origin (with child camera) for AR mode. We then flip between the two cameras when the user toggles viewing modes.

Another issue you may run into is determining whether the user's device supports a specific AR feature at runtime.

Determining whether the device supports a specific AR feature

It is possible that your app requires a specific AR feature that is not supported by all devices. We can ask the Unity AR subsystems what features are supported by getting the subsystem descriptor records.

For example, suppose we are interested in detecting vertical planes. Some older devices may support AR but only horizontal planes. The following code illustrates how to get and check plane detection support:

```
using System.Collections.Generic;
using UnityEngine;
using UnityEngine.XR.ARSubsystems;


public class CheckPlaneDetectionSupport : MonoBehaviour
{
    void Start()
    {
        var planeDescriptors =
            new List<XRPlaneSubsystemDescriptor>();
        SubsystemManager.
            GetSubsystemDescriptors(planeDescriptors);

        Debug.Log("Plane descriptors count: " +
            planeDescriptors.Count);

        if (planeDescriptors.Count > 0)
        {
            foreach (var planeDescriptor in planeDescriptors)
            {
                Debug.Log("Support horizontal: " +
                    planeDescriptor.
                        supportsHorizontalPlaneDetection);
                Debug.Log("Support vertical: " +
                    planeDescriptor.
                        supportsVerticalPlaneDetection);
                Debug.Log("Support arbitrary: " +
                    planeDescriptor.
                        supportsArbitraryPlaneDetection);
```

```
                Debug.Log("Support classification: " +
                    planeDescriptor.supportsClassification);
        }
    }
  }
}
```

The types of descriptors available in AR Foundation include the following (their purpose is self-evident from their names):

- XRPlaneSubsystemDescriptor
- XRRaycastSubsystemDescriptor
- XRFaceSubsystemDescriptor
- XRImageTrackingSubsystemDescriptor
- XREnvironmentProbeSubsystemDescriptor
- XRAnchorSubsystemDescriptor
- XRObjectTrackingSubsystemDescriptor
- XRParticipantSubsystemDescriptor
- XRDepthSubsystemDescriptor
- XROcclusionSubsystemDescriptor
- XRCameraSubsystemDescriptor
- XRSessionSubsystemDescriptor
- XRHumanBodySubsystemDescriptor

Documentation for the AR Subsystems API and these descriptor records can be found at https://docs.unity3d.com/Packages/com.unity.xr.arsubsystems@4.2/api/UnityEngine.XR.ARSubsystems.html. For example, the XRPlaneSubsystemDescriptor record we used here is documented at https://docs.unity3d.com/Packages/com.unity.xr.arsubsystems@4.2/api/UnityEngine.XR.ARSubsystems.XRPlaneSubsystemDescriptor.Cinfo.html.

If you are planning to distribute your application in different countries, you may also be interested in localization.

Adding localization

Localization is the translation of text strings and other assets into local languages. It can also specify date and currency formatting, alternative graphics for national flags, and so on, to accommodate international markets and users. The Unity Localization package provides a standard set of tools and data structures for localizing your application. More information can be found at `https://docs.unity3d.com/Packages/com.unity.localization@0.10/manual/QuickStartGuide.html`. We do not use localization in any projects in this book, except where already supported by imported assets such as the Onboarding UX assets from the AR Foundation Demos project.

The Unity Onboarding UX assets has built-in support for localization of the user prompts and explanation of scanning problems. The `ReasonsUX` localization tables given with the Onboarding UX project, for example, can be opened by selecting **Window | Asset Management | Localization Tables** and is shown in the following screenshot. You can see, for example, the second-row **INIT** key says in English, **Initializing augmented reality**, along with the same string translated into many other languages:

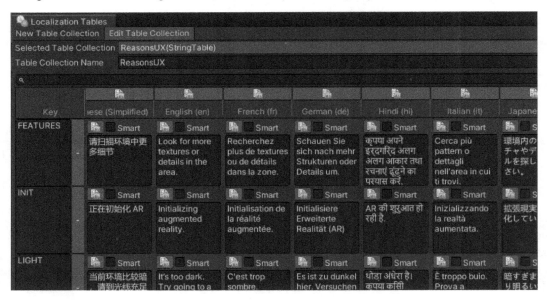

Figure 5.11 – The ReasonsUX localization tables included in Onboarding UX assets

In the code, the **Initializing augmented reality** message, for example, is retrieved with a call like this:

```
string localizedInit = reasonsTable.GetEntry("INIT").
GetLocalizedString();
```

When we added the onboarding UX prefab (`ARFoundationDemos/UX/Prefabs/ScreenspaceUI`) to our scene, I had you disable the **Localization Manager** component because it gives runtime errors until it is set up. Provided you've installed the **Localization** package via **Package Manager** as described earlier in this chapter, we can set it up now for the project using the following steps:

1. Open the **Localization** settings window by going to **Edit | Project Settings | Localization**.

2. In the **Project** window, navigate to `Assets/ARFOundationDemos/Common/Localization/` and drag the `LocalizationSettings` asset onto the **Location Settings** slot (or use the *doughnut* icon to open the **Location Setting Select** dialog box).

3. In the settings window, click **Add All**.

4. In the **Hierarchy** window, select the **OnboardingUX** object and in the **Inspector**, enable the **Localization Manager** component.

5. Open the **Addressables Groups** window using **Window | Asset Management | Addressables | Groups**.

6. From the **Addressables Groups** menu bar, select **Build | New Build | Default Build Script**. You will need to do this for each target platform you are building (for example, once for Android and once for iOS).

As you can see in this last step, the Localization package uses Unity's new **Addressables** system for managing, packing, and loading assets from any location locally or over the internet (`https://docs.unity3d.com/Packages/com.unity.addressables@1.12/manual/index.html`).

Note that as I'm writing this, the Onboarding UX `LocalizationManager` script does not select the language at runtime. The language must be set in the Inspector and compiled into your build.

The AR UI framework we built in this chapter can be used as a template for new scenes. Unity makes it easy to set that up.

Summary

In this chapter, we got a chance to use the AR user framework we developed in the previous *Chapter 4, Creating an AR User Framework*, in a simple AR *Place Object Demo* project. We created a new scene using the ARFramework scene template that implements a state machine mechanism for managing user interaction modes. It handles user interaction with a controller-view design pattern, separating the control scripts from the UI graphics.

By default, the scene includes the AR Session and AR Session Origin components required by AR Foundation. The scene is set up with a Canvas UI containing separate panels that will be displayed for each interaction mode. It also includes an Interaction Controller that references separate mode objects, one for each interaction mode.

The modes (and corresponding UI) given with the template are Startup, Scan, Main, and NonAR. An app using this framework first starts in Startup-mode while the AR Session is initializing. Then it goes into Scan-mode, prompting the user to scan the environment for trackable features, until a horizontal plane is detected. Then it goes into Main-mode and displays the main menu.

For this project, we added a main menu that is displayed during Main-mode and that contains buttons for placing various virtual objects in the environment. Pressing a button enables a new PlaceObject-mode that we added to the scene. When PlaceObject-mode is enabled, it displays an instructional animated prompt for the user to tap to place an object in the scene. After an object is added, the app returns to Main-mode, and the trackables are hidden so you can see your virtual objects in the real world without any extra distractions.

In the next chapter, we will go beyond a simple demo project and begin to build a more complete AR application – a photo gallery where you can place framed photos of your favorite pictures on the drab walls in your home or office.

Section 3 – Building More AR Projects

In this section, we will build a variety of AR projects, each using different feature detection techniques supported by AR Foundation, including plane detection, image recognition, and face tracking. The result of each project is a working demo that could be improved and developed into a more complete working app.

This section comprises the following chapters:

- *Chapter 6, Gallery: Building an AR App*
- *Chapter 7, Gallery: Editing Virtual Objects*
- *Chapter 8, Planets: Tracking Images*
- *Chapter 9, Selfies: Making Funny Faces*

6
Gallery: Building an AR App

In this chapter, we will begin building a full **Augmented Reality** (**AR**) app, an AR *art gallery* that lets you hang virtual framed photos on your real-world walls.

First, we'll define the goals of the project and discuss the importance of project planning and **user experience** (**UX**) design. When the user presses the **Add** button in the main menu, they'll see a **Select Image** menu. When they pick one, they'll be prompted to place a framed copy of the image on their real-world wall.

To implement the project, we will start with the AR user framework scene template that we created earlier in this book. We'll build a Select Image UI panel and interaction mode, and define the image data used by the app.

In this chapter, we will cover the following topics:

- Specifying a new project and UX design
- Using data structures and arrays, and passing data between objects
- Creating a detailed UI menu panel with a grid of buttons
- Creating prefabs for instantiating in an AR scene
- Implementing a complete scenario based on a given user story

By the end of the chapter, you'll have a working prototype of the app that implements one scenario: placing pictures on the wall. Then we'll continue to build and improve the project in the next chapter.

Technical requirements

To implement the project in this chapter, you need Unity installed on your development computer, connected to a mobile device that supports AR applications (see *Chapter 1, Setting Up for AR Development,* for instructions). We also assume that you have the `ARFramework` template and its prerequisites installed; see *Chapter 5, Using the AR User Framework*. The completed project can be found in this book's GitHub repository, `https://github.com/PacktPublishing/Augmented-Reality-with-Unity-AR-Foundation`.

Specifying the Art Gallery project UX

An important step before beginning any new project is to do some design and specifications ahead of time. This often entails writing it down in a document. For games, this may be referred to as the **Game Design Document** (GDD). For applications, it may be a **Software Design Document** (SDD). Whatever you call it, the purpose is to put into writing a blueprint of the project before development begins. A thorough design document for a Unity AR project might include details such as the following:

- *Project overview*: Summarize the concept and purpose of the project, identify the primary audience, and perhaps include some background on why the project exists and how and why it will be successful.

- *Use cases*: Identify the real-life problems the product will solve. It's often effective to define separate user **personas** (with real or fictitious names) representing types of users of the application, their main goals, and how they'll use the app to achieve these objectives.

- *Key features*: Identify the discrete areas of functionality that deliver value to your users, perhaps with an emphasis on what distinguishes it from other similar solutions.

- *UX design*: The **user experience (UX)** design may include a variety of user scenarios that detail specific workflows, often presented as a **storyboard** using abstract pencil or wireframe sketches. In lieu of drawing skills, photo captures of a whiteboard session and sticky notes may be sufficient.

  Separately, you may also include UI graphic designs that define actual style guides and graphics, for example, color schemes, typography, button graphics, and so on.

- *Assets*: Collect and categorize the graphic assets you anticipate needing, including concept art, 3D models, effects, and audio.

- *Technical plan*: This includes software architecture and design patterns that will be used, development tools (such as Unity, Visual Studio, and GitHub), the Unity version, third-party packages (for example, via Package Manager), plus Unity Services and other cloud services (such as advertising, networking, and data storage).

- *Project plan*: The implementation plan may show the anticipated project phases, production, and release schedules. This could involve the use of tools such as Jira or Trello.

- *Business plan*: Non-technical planning may include plans for project management, marketing, funding, monetization, user acquisition, and community-building.

For very large projects, these sections could be separate documents. For small projects, the entire thing may only be a few pages long with bullet points. Just keep in mind that the main purpose is to think through your plans before committing to code. That said, don't over-design. Keep in mind one of my favorite quotes from Albert Einstein:

"Make everything as simple as possible, but not simpler."

Assume things can and will change as the project progresses. Rapid iteration, frequent feedback from stakeholders, and engaging real users may reaffirm your plans. Or it may expose serious shortcomings with an original design and can take a project in new, better directions. As I tell my clients and students:

"The time you know least about a project is at the beginning!"

In this book, I'll provide an abbreviated design plan at the beginning of each project that tries to capture the most important points without going into a lot of detail. Let's start with this AR Gallery project, and spec out the project objective, use cases, a UX design, and a set of user stories that define the key features of the project.

Project objectives

We are going to build an AR art gallery project that allows users to place their favorite photos on walls of their home or office as virtual framed images using AR.

Use cases

Persona: Jack. Jack works from home and doesn't have time to decorate his drab apartment. Jack wants to spruce up the walls by adding some nice pictures on the wall. But his landlord doesn't allow putting nails in the walls. John also wants to be able to change his hung pictures frequently. Jack spends many hours per day using his mobile phone, so looking at the walls through his phone is satisfying.

Persona: Jill. Jill has a large collection of favorite photos. She would like to hang them on the walls of her office but it's not very appropriate for a work environment. Also, she is a bit obsessive and thus would like to frequently rearrange the photos and swap the pictures.

UX design

The **user experience (UX)** for this application must include the following requirements and scenarios:

- When the user wants to place a photo on the wall, they select an image from a menu and then tap the screen, indicating where to place the photo.

- When the user wants to modify a photo already placed on the wall, they can tap the photo to enable editing. Then the user can drag to move, pinch to resize, choose a different photo or frame, or swipe to remove the photo.

- When the framed photo is rendered, it matches the current room lighting conditions and casts shadows on real-world surfaces.

- When the user exits and re-opens the app, all the photos they placed in the room will be saved and restored in their locations.

I asked a professional UX designer (and friend of mine) Kirk Membry (https://kirkmembry.com/) to prepare UX wireframe sketches specifically for this book's project. The following image shows a few frames of a full storyboard:

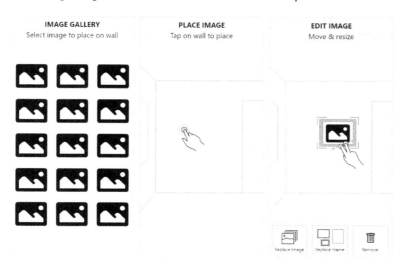

Figure 6.1 – UX design wireframe sketches

The leftmost frame shows the image gallery menu that appears when the user has chosen to add a new photo into the scene. The middle frame depicts the user choosing a location to hang the photo on a wall. And the rightmost frame shows the user editing an existing picture, including finger gestures to move and resize, and a menu of other edit options on the bottom of the screen.

Storyboards like this can be used to communicate the design intent to graphic designers, coders, and stakeholders alike. It can form the basis of discussion for ironing out kinks in the user workflow and inconsistencies in the user interface. It can go a long way to make the project management more efficient by preventing unnecessary rework when it's most costly – after features have been implemented.

With enough of the design drafted, we can now select some of the assets we'll use while building the project.

User stories

It is useful to break up the features into a set of "user stories" or bite-sized features that can be implemented incrementally, building up the project a piece at a time. In an agile-managed project, the team may choose a specific set of stories to accomplish in one- or two-week *sprints*. And these stories could be managed and tracked on a shared project board such as Trello (`https://trello.com/`) or Jira (`https://www.atlassian.com/software/jira`). Here are a set of stories for this project:

- When the app starts, I am prompted to scan the room while the device detects and tracks vertical walls in the environment.

- After tracking is established, I see a main menu with an **Add** button.

- When I press the **Add** button, I am presented with a selection of photos.

- When I choose a photo from the selection, I see the tracked vertical planes and I am prompted to tap to hang a framed photo (picture) on a wall.

- When the picture is instantiated, it hangs squarely upright and flush against the wall plane.

- When I tap on an existing virtual picture to begin editing the picture.

- When editing a picture, I see an edit menu with options to change the photo, change the frame, or remove the framed picture.

- When editing a picture, I can drag the picture to a new location.

- When editing a picture, I can pinch (using two fingers) to resize it.

That seems like a good set of features. We'll try to get through the first half of them in this chapter and complete it in the next chapter. Let's get started.

Getting started

To begin, we'll create a new scene named `ARGallery` using the `ARFramework` scene template, with the following steps:

1. Select **File | New Scene**.

2. In the **New Scene** dialog box, select the **ARFramework** template.

3. Select **Create**.

4. Select **File | Save As**. Navigate to the `Scenes/` folder in your project's `Assets` folder, give it the name `ARGallery`, and select **Save**.

The new AR scene already has the following objects:

- An **AR Session** game object.

- An **AR Session Origin** rig with raycast manager and plane manager components.

- **UI Canvas** is a screen space canvas with child panels Startup UI, Scan UI, Main UI, and NonAR UI. It has the UI Controller component script that we wrote.

- **Interaction Controller** is a game object with the Interaction Controller component script we wrote that helps the app switch between interaction modes, including Startup, Scan, Main, and NonAR modes. It also has a **Player Input** component configured with the **AR Input Actions** asset we created previously.

- An **OnboardingUX** prefab from the AR Foundation Demos project that provides AR session status and feature detection status messages, and animated onboarding graphics prompts.

We now have a plan for the AR gallery project, including a statement of objectives, use cases, and a UX design with some user stories to implement. With this scene, we're ready to go. Let's find a collection of photos we can work with and add them to the project.

Collecting image data

In Unity, images can be imported for use in a variety of purposes. Textures are images that can be used for texturing the materials for rendering the surface of 3D objects. The UI uses images as sprites for button and panel graphics. For our framed photos, we're going to use images as… images.

The most basic approach to using images in your application is to import them into your `Assets` folder and reference them as Unity textures. A more advanced solution would be to dynamically find and load them at runtime. In this chapter, we'll use the former technique and build the list of images into the application. Let's start by importing the photos you want to use.

Importing photos to use

Go ahead and choose some images for your gallery from your favorites. Or you can use the images included with the files in this book's GitHub repository, containing a collection of freely usable nature photos found on Unsplash.com (`https://unsplash.com/`) that I found, along with a photo of my own named `WinterBarn.jpg`.

To import images into your project, use the following steps:

1. In the **Project** window, create a folder named `Photos` by right-clicking, then selecting **Create | Folder**.

2. From your Windows Explorer or OSX Finder, locate an image you want to use. Then drag the image file into Unity, dropping it in your `Photos/` folder.

3. In the **Inspector** window, you can check the size of the imported image. Because we're using it in AR and on a relatively low-resolution mobile device, let's limit the max size to 1,024 pixels. Note that Unity requires textures be imported into a size that is a power of 2 for best compression and runtime optimization. In the **Inspector**, ensure the **Default** tab is selected and choose **Max Size | 1024**.

Now we'll add a way to reference your images in the scene.

Adding image data to the scene

To add the image data to the scene, we'll create an empty GameObject with an `ImagesData` script that contains a list of images. First, create a new C# script in your project's `Scripts/` folder, name it `ImagesData`, and write it as follows:

```
using UnityEngine;


[System.Serializable]
public struct ImageInfo
{
    public Texture texture;
    public int width;
    public int height;
}
public class ImagesData : MonoBehaviour
{
    public ImageInfo[] images;
}
```

The script starts by defining an `ImageInfo` data structure containing the image `Texture` and the pixel dimensions of the image. It is `public` so it can be referenced from other scripts. Then the `ImagesData` class declares an array of this data in the `images` variable. The `ImageInfo` structure requires a `[System.Serializable]` directive so it will appear in the Unity Inspector.

Now we can add the image data to the scene, using the following steps:

1. From the main menu, select **GameObject | Create Empty** to add an object to the root of your hierarchy, and rename it `Images Data` (reset its **Transform** for tidiness, using the 3-dot context menu and **Reset**).

2. Drag the **ImagesData** script onto the **Images Data** object, making it a component.

3. To populate the `images` array, in the **Inspector**, enter the number of images you plan to use, or simply press the + button in the bottom right to incrementally add elements to the array.

4. Add your imported image files one at a time by unfolding an **Element** from the **Images** list, then drag an image file from the **Project** window onto the **Texture** slot for the element. Please also enter the **Width** and **Height** in pixels of each image.

   My **Images Data** looks like this in the **Inspector**:

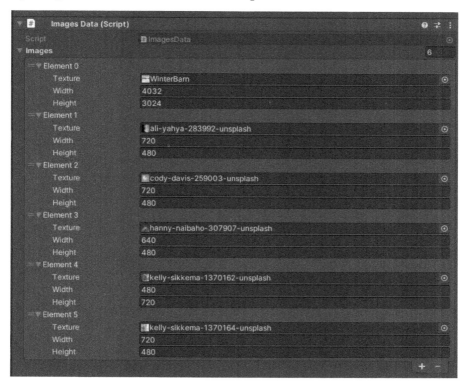

Figure 6.2 – Images Data component with a list of images

> **Using ScriptableObjects**
>
> A different, and probably better, approach to providing the list of images is
> to use ScriptableObjects instead of GameObjects. ScriptableObjects are data
> container objects that live in your `Assets/` folder rather than in the scene
> hierarchy. You can learn more about ScriptableObjects at `https://docs.`
> `unity3d.com/Manual/class-ScriptableObject.html` and
> `https://learn.unity.com/tutorial/introduction-to-`
> `scriptable-objects`.

It is a little tedious having to manually enter the pixel dimensions of each image. It would
be nice if there were a better way because that's not very easy.

Obtaining the pixel dimensions of an image

Unfortunately, when Unity imports an image as a texture, it resizes it to a power of two to
optimize runtime performance and compression, and the original dimension data is not
preserved. There are several ways around this, none of which are very pretty:

- Require the developer to specify the pixel dimensions manually for each image. This
  is the approach we are taking here.

- Tell Unity to not resize the image when it is imported. For this, select an image
  asset, and in its **Inspector** window, you'll see its **Import Settings**. Notice its physical
  size on disk is listed in the preview panel at the bottom. Then change **Advanced
  | Non-Power of 2** to **None** and select **Apply**. Note the new size is probably
  significantly bigger because Unity will not compress the data. And that will
  make your final app size much bigger too. But since the texture is now the
  original unscaled size, you can access it in C# using `Texture.width` and
  `Texture.height`.

- Take the first method but automatically determine the pixel size using an Editor
  script. Unity allows you to write scripts that only run in the Editor, not runtime.
  The Editor has access to the original image file in your **Assets** folder before it has
  been imported as a texture. So it's possible to read and query this information,
  either using system I/O functions, or possibly (undocumented) the Unity API (see
  `https://forum.unity.com/threads/getting-original-size-of-`
  `texture-asset-in-pixels.165295/`).

Given that, we'll stick with the manual approach in this chapter, and you can explore the other options on your own.

Perhaps you're also wondering, what if I don't want to build the images into my project and want to find and load them at runtime?

Loading the pictures list at runtime

Loading assets at runtime from outside your build is an advanced topic and outside the scope of this chapter. There are several different approaches that I will briefly describe, and I will point you to more information:

- **Including images in Asset Bundles**: In Unity, you have the option of bundling assets into an Asset Bundle that the application can download after the user has installed the app, as **downloadable content** (**DLC**). See `https://docs. unity3d.com/Manual/AssetBundlesIntro.html`.

- **Downloading images from a web URL**: If you have the web address of an image file, you can download the image at runtime using a web request and use it as a texture in the app. See `https://docs.unity3d.com/ScriptReference/ Networking.UnityWebRequestTexture.GetTexture.html`.

- **Getting images from the device's photos app**: For an application such as our Gallery, it's natural to want to get photos from the user's own photos app. To access data from other apps on the mobile device you need a library with native access. It may also require your app to obtain additional permissions from the user. Search the Unity Asset Store for packages.

If you want to implement these features, I'll leave that up to you.

We have now imported the photos we plan to use, created a C# `ImageInfo` data structure including the pixel dimensions of each image, and populated this image data in the scene. Let's create a framed photo prefab containing a default image and a picture frame that we can place on a wall plane.

Creating a framed photo prefab

The user will be placing a framed photo on their walls. So, we need to create a prefab game object that will be instantiated. We want to make it easy to change images and frames, as well as resize them for various orientations (landscape versus portrait) and image aspect ratios. For the default frame, we'll create a simple block from a flattened 3D cube and mount the photo on the face of it. For the default image, you may choose your own or use one that's included with the files for this chapter in the GitHub repository.

Creating the prefab hierarchy

First, create an empty prefab named `FramedPhoto` in your project's `Assets/` folder. Follow these steps:

1. In the **Project** window, navigate to your `Prefabs/` folder (create one if needed). Then *right-click* in the folder and select **Create | Prefab**.

2. Rename the new prefab `FramedPhoto`.

3. *Double-click* the **FramedPhoto** asset (or click its **Open Prefab** button in the **Inspector** window).

   We're now editing the empty prefab.

4. Add a child **AspectScaler** container that we can later use to adjust its aspect ratio for the given image: *right-click* in the **Hierarchy** window and select **Create Empty** (or use the + button in the top-left of the window). Rename it `AspectScaler`.

5. Let's create a modern-looking rectangular black frame using a flattened cube. With the **AspectScaler** object selected, *right-click* and select **3D Object | Cube** and rename it `Frame`.

6. Give the frame some thickness. In the frame's **Inspector** window, set its **Transform | Scale | Z** to `0.05` (that's in meters).

7. Likewise, offset it from the wall by setting **Transform | Position | Z** to `-0.025`.

8. To give this frame a black finish, create and add a new material as follows.

   In the **Project** window, navigate to your `Materials/` folder (create one if needed). Then *right-click* in the folder and select **Create | Material**. Rename the new material `Black Frame Material`.

9. Set its **Base Map** color to a charcoal black color.

10. Then, in the **Hierarchy**, select the **Default Frame** object and drag the **Black Frame Material** onto it.

The current frame properties are shown in the following screenshot:

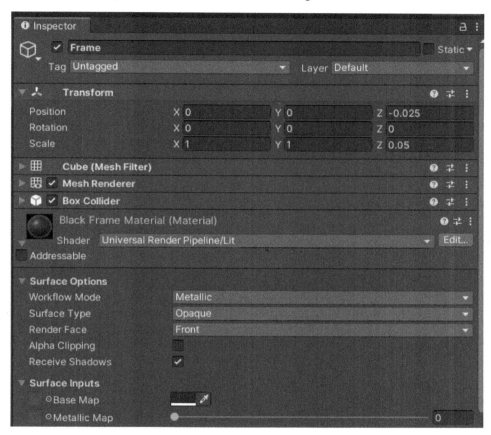

Figure 6.3 – The FramedPhoto's frame properties

Next, we'll add a default image to the **FramedPhoto** rig. I'm using the one named WinterBarn.jpg that is included with the files for this book. Use the following steps to create an image object with a material that uses this photo as its texture image:

1. With the **FramedPhoto** prefab open for editing, in **Hierarchy**, *right-click* on the **AspectScaler** object, select **Create | 3D Object | Quad**, and rename it Image. A **quad** is the simplest Unity 3D primitive object, a flat quadrilateral plane with only four edges and facing in a single direction.

2. To add your image as a texture on the quad, we need to create a material. In the **Project** window, navigate to your `Materials/` folder, *right-click* in the folder and select **Create | Material**. Rename the new material `Image Material`.

3. Drag your image file (`WinterBarn.jpg`) from the **Project** window into the **Inspector** window, dropping it onto the little square "chip" slot on the left side of the **Base Map** property.

4. Drag the **Image Material** onto the **Image** game object.

5. Offset the image quad so it's slightly in front of the frame cube's plane. Set its **Transform | Position | Z** to `-0.06`.

6. You should be able to see the image now. But the frame is hidden because the image quad is scaled to the same size as the frame. Shrink the image by setting its **Scale X** and **Y** to `0.9`.

The prefab hierarchy now looks like the following screenshot, where the image is currently selected and visible in the **Inspector**:

Figure 6.4 – The FramedPhoto prefab

Next, let's add a simple script that will help our other code set the image of a **FramedPhoto** object.

Writing a FramedPhoto script

We are going to need to set various properties of each instance of the **FramedPhoto** prefab. Specifically, the user will be able to choose which image belongs in the frame of each picture. So, we can provide a `SetImage` function for this that gets the image data for this picture.

Create a new C# script named `FramedPhoto`, open it for editing, and write the script as follows::

```
using UnityEngine;

public class FramedPhoto : MonoBehaviour
{
    [SerializeField] Transform scalerObject;
    [SerializeField] GameObject imageObject;

    ImageInfo imageInfo;

    public void SetImage(ImageInfo image)
    {
        imageInfo = image;

        Renderer renderer =
            imageObject.GetComponent<Renderer>();
        Material material = renderer.material;
        material.SetTexture("_BaseMap", imageInfo.texture);
    }
}
```

At the top of the `FramedPhoto` class, we declare two properties. The `imageObject` is a reference to the child **Image** object, for when the script needs to set its image texture. `scalerObject` is a reference to the **AspectScaler** for when the script needs to change its aspect ratio (we do this at the end of this chapter).

When a **FramedPhoto** gets instantiated, we are going to call `SetImage` to change the **Image** texture to the one that should be displayed. The code required to do this takes a few steps. If you look at the **Image** object in the Unity Inspector, you can see it has a **Renderer** component that references its **Material** component. Our script gets the `Renderer`, then gets its `Material`, and then sets its base texture.

We can now add this script to the prefab as follows:

1. With the **FramedPhoto** prefab opened for editing, drag the **FramedPhoto** script onto the **FramedPhoto** root object to make it a component.

2. From the **Hierarchy**, drag the **AspectScaler** object into the **Inspector** and drop it onto the **Framed Photo | Scaler Object** slot.

3. From the **Hierarchy**, drag the **Image** object onto the **Framed Photo | Image Object** slot.

Our prefab is now almost ready to be used. Of course, the picture we're using isn't really supposed to be square, so let's scale it.

Scaling the picture's shape

The photo I'm using by default is landscape orientation, but our frame is square, so it looks squished. To fix it, we need to get the original pixel size of the image and calculate its aspect ratio. For example, the `WinterBarn.jpg` image included on GitHub for this book is 4,032x3,024 (width x height), or 3:4 (`height:width` landscape ratio). Let's scale it now for the image's aspect ratio (`0.75`). Follow these steps:

1. In the **Hierarchy** window, select the **Scaler** object.

2. Set its **Transform | Scale Y** to `0.75` (if your image is portrait, scale the **X** axis instead, leaving the **Y** axis at `1.0`).

   The properly scaled prefab now looks like the following:

Figure 6.5 – FramedPhoto prefab with corrected 3:4 landscape aspect ratio

3. Save the prefab by clicking the **Save** button in the top-right of the **Scene** window.

4. Return to the scene editor using the < button in the top-left of the **Hierarchy** window.

5. Setting up the **FramedPhoto** rig this way has advantages, including the following:

- The **FramedPhoto** prefab is normalized to unit scale, that is, scaled (`1, 1, 1`) regardless of the aspect ratio of the photo within it or the thickness of the frame. This will help with the user interface for placing and scaling the framed photo in the scene.

- The **FramedPhoto** prefab's anchor point is located at the center of the picture along the back face of the frame, so when it's placed on a wall it'll be positioned flush with the detected wall plane.

- The **Frame** model and photo **Image** objects are within an **AspectScaler** object that can be scaled according to the aspect ratio of the image. By default, we set it to `0.75` height for the 3:4 aspect ratio.

- The **Image** is scaled evenly (that is, by the same ratio for both X and Y) to fit within the picture area of the frame. In this case, I decided the frame has a `0.05` size border, so the **Image** is scaled by `0.9`.

- The front-back offset of the image will also depend on the frame's model. In this case, I moved it closer, `-0.06` versus `-0.025` units, so it sits slightly in front of the frame's surface.

When assembling a prefab, thinking through how it can head off gotchas later.

In this section, we created a scalable **FramedPhoto** prefab made from a cube and an image mounted on the face of the frame block that we can now add to our scene. It is saved in the project `Assets` folder so copies can be instantiated in the scene when the user places a picture on a wall. The prefab includes a `FramedPhoto` script that manages some aspects of the behavior of the prefab, including setting its image texture. This script will be expanded later in the chapter. We now have a **FramedPhoto** prefab with a frame. We're ready to add the user interaction for placing pictures on your walls.

Hanging a virtual photo on your wall

For this project, the app scans the environment for vertical planes. When the user wants to hang a picture on the wall, we'll show a UI panel that instructs the user to tap to place the object, using an animated graphic. Once the user taps the screen, the **AddPicture** mode instantiates a **FramedPhoto** prefab, so it appears to hang on the wall, upright and flush against the wall plane. Many of these steps are similar to what we did in *Chapter 5, Using the AR User Framework*, so I'll offer a little less explanation here. We'll start with a similar script and then enhance it.

Detecting vertical planes

Given the AR Session Origin already has an AR Plane Manager component (provided in the default `ARFramework` template), use the following steps to set up the scene to scan for vertical planes (instead of horizontal ones):

1. In the **Hierarchy** window, select the **AR Session Origin** object.

2. In its **Inspector** window, set the **AR Plane Manager | Detection Mode** to **Vertical** by first selecting **Nothing** (clearing all the selections) and then selecting **Vertical**.

Now let's create the **AddPicture** UI panel that prompts the user to tap a vertical plane to place a new picture.

Creating the AddPicture UI panel

The **AddPicture UI** panel is similar to the **Scan UI** one included with the scene template, so we can duplicate and modify it as follows:

1. In the **Hierarchy** window, unfold the **UI Canvas**.

2. *Right-click* the **Scan UI** game object and select **Duplicate**. Rename the new object `AddPicture UI`.

3. Unfold **AddPicture UI** and select its child, **Animated Prompt**.

4. In the **Inspector**, set the **Animated Prompt | Instruction** to **Tap To Place**.

5. To add the panel to the UI Controller, in the **Hierarchy**, select the **UI Canvas** object.

6. In the **Inspector**, at the bottom-right of the **UI Controller** component, click the **+** button to add an item to the UI Panels dictionary.

7. Enter `AddPicture` in the **Id** field.

8. Drag the **AddPicture UI** game object from the **Hierarchy** onto the **Value** slot.

We added an instructional user prompt for the **AddPicture** UI. When the user chooses to add a picture to the scene, we'll go into **AddPicture** mode, and this panel will be displayed. Let's create the **AddPicture** mode now.

Writing the initial AddPictureMode script

To add a mode to the framework, we create a child GameObject under the **Interaction Controller** and write a mode script. The mode script will show the mode's UI, handle any user interactions, and then transition to another mode when it is done. For AddPicture mode, it will display the **AddPicture UI** panel, wait for the user to tap the screen, instantiate the prefab object, and then return to main mode.

The script starts out like the `PlaceObjectMode` script we wrote in *Chapter 5, Using the AR User Framework*. Then we'll enhance it to ensure the framed picture object is aligned with the wall plane, facing into the room, and hanging straight.

Let's write the `AddPictureMode` script, as follows:

1. Begin by creating a new script in your project's `Scripts/` folder by right-clicking and selecting **Create C# Script**. Name the script `AddPictureMode`.

2. *Double-click* the file to open it for editing. Paste the following code, which is the same as the **PlaceObjectMode** script you may already have to hand, with differences highlighted. The first half of the script is as follows:

```csharp
using System.Collections.Generic;
using UnityEngine;
using UnityEngine.InputSystem;
using UnityEngine.XR.ARFoundation;
using UnityEngine.XR.ARSubsystems;

public class AddPictureMode : MonoBehaviour
{
    [SerializeField] ARRaycastManager raycaster;
    [SerializeField] GameObject placedPrefab;
    List<ARRaycastHit> hits = new List<ARRaycastHit>();

    void OnEnable()
    {
        UIController.ShowUI("AddPicture");
    }
}
```

3. The second part of the script is actually unchanged from the `PlaceObjectMode` script:

```
public void OnPlaceObject(InputValue value)
{
    Vector2 touchPosition = value.Get<Vector2>();
    PlaceObject(touchPosition);
}

void PlaceObject(Vector2 touchPosition)
{
    if (raycaster.Raycast(touchPosition, hits,
        TrackableType.PlaneWithinPolygon))
    {
        Pose hitPose = hits[0].pose;
        Instantiate(placedPrefab, hitPose.position,
            hitPose.rotation);

        InteractionController.EnableMode("Main");
    }
}
}
```

At the top of `AddPictureMode`, we declare a `placedPrefab` variable that will reference the **FramedPhoto Prefab** asset we created. We also define and initialize references to the `ARRaycastManager` and a private list of `ARRaycaseHit hits` that we'll use in the `PlaceObject` function.

When the mode is enabled, we show the `AddPicture` UI panel. Then, when there's an `OnPlaceObject` user input action event, `PlaceObject` does a `Raycast` on the trackable planes. If there's a hit, it instantiates a copy of the **FramedPhoto** into the scene, and then goes back to main mode.

Let's go with this initial script for now and fix any problems we discover later. The next step is to add the AddPicture mode to the app.

Creating the AddPicture Mode object

We can now add the AddPicture mode to the scene by creating an **AddPicture Mode** object under the **Interaction Controller**, as follows:

1. In the **Hierarchy** window, *right-click* the **Interaction Controller** game object and select **Create Empty**. Rename the new object `AddPicture Mode`.

2. Drag the `AddPictureMode` script from the **Project** window onto the **AddPicture Mode** object, adding it as a component.

3. Drag the **AR Session Origin** object from the **Hierarchy** onto the **Add Picture Mode | Raycaster** slot.

4. Locate your **FramedPhoto Prefab** asset in the **Project** window and drag it onto the **Add Picture Mode | Placed Prefab** slot. The **AddPicture Mode** component now looks like the following (note that this screenshot also shows two more parameters, **Image Data** and **Default Scale**, that we add to the script at the end of this chapter):

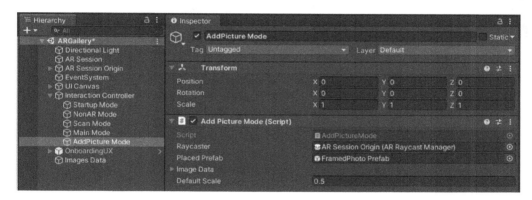

Figure 6.6 – AddPicture Mode added to the scene

5. Now we'll add the mode to the **Interaction Controller**. In the **Hierarchy**, select the **Interaction Controller** object.

6. In the **Inspector**, at the bottom-right of the **Interaction Controller** component, click the + button to add an item to the **Interaction Modes** dictionary.

7. Enter `AddPicture` in the **Id** field.

8. Drag the **AddPicture Mode** game object from the **Hierarchy** onto the **Value** slot. The Interaction Controller component now looks like the following:

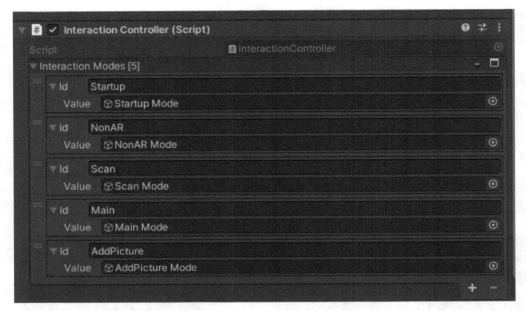

Figure 6.7 – Interaction Controller with AddPicture Mode added to the Interaction Modes dictionary

We now have an **AddPicture** mode that will be enabled from **Main** mode when the user clicks an **Add** button. Let's create this button now.

Creating a main menu Add button

When the app is in Main mode, the **Main UI** panel is displayed. On this panel, we'll have an **Add** button for the user to press when they want to place a new picture in the scene. I'll use a large plus sign as its icon, with the following steps:

1. In the **Hierarchy** window, unfold the **UI Canvas** object, and unfold its child **Main UI** object.

2. The default child text in the panel is a temporary placeholder; we can remove it. *Right-click* the child **Text** object and select **Delete**.

3. Now we add a button. *Right-click* the **Main UI** game object and select **UI | Button – TextMeshPro**. Rename it Add Button.

4. With the **Add Button** selected, in its **Inspector** window use the *anchor menu* (upper-left) to select a **Bottom-Right** anchor. Then press *Shift + Alt + click* **Bottom-Right** to also set its **Pivot** and **Position** in that corner.

5. Adjust the button size and position, either using the *Rect Tool* from the **Scene** window toolbar (the fifth icon from the left) or numerically in the Inspector, such as **Width, Height** to (175, 175), and **Pos X, Pos Y** to (-30, 30), as shown in the following screenshot:

Figure 6.8 – The Add button Rect Transform settings

6. In the **Hierarchy** window, unfold the **Add Button** by clicking its *triangle-icon* and select its child object, **Text (TMP)**.

7. Set its **Text** value to + and set its **Font Size** to 192.

8. You can add another text element to label the button. *Right-click* the **Add Button** and select **UI | Text – TextMeshPro**. Set its **Text** content to Add, **Font Size**: 24, **Color**: black, **Alignment**: center, and **Rect Transform | Pos Y** to 55.

Our button now looks like the following:

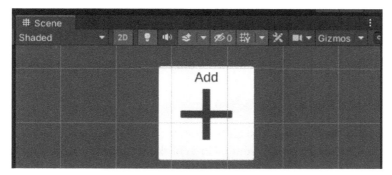

Figure 6.9 – The Add button

9. To set up the button to enable **PlacePicture Mode**, select the **Add Button** in the **Hierarchy**. In its **Inspector**, in the **OnClick** section of the **Button** component, press the + button on the bottom-right to add an event action.

10. Drag the **Interaction Controller** from the **Hierarchy** and drop it onto the **Object** slot of the **OnClick** action.

11. In the **Function** select list, choose **InteractionController | EnableMode**.

12. In its string parameter field, enter the text AddPicture.

The **On Click** property of now looks like this:

Figure 6.10 – When the Add button is clicked, it calls EnableMode("AddPicture")

We have now added **AddPicture Mode** to our framework. It will be enabled by the Interaction Controller when the **Add** button is clicked. When enabled, the script shows the **AddPicture** instructional UI, then waits for a **PlaceObject** input action event. Then it uses Raycast to determine where in 3D space the user wants to place the object, instantiates the prefab, and then returns to Main mode. Let's try it out.

Build And Run

Save the scene. If you want to try and see how it looks, you can now **Build And Run**, as follows:

1. Select **File | Build Settings**.

2. Click the **Add Open Scenes** button if the current scene (ARGallery) is not already in the **Scenes In Build** list.

3. Ensure that the ARGallery scene is the only one checked in the **Scenes In Build** list.

4. Click **Build And Run** to build the project.

The app will start and prompt you to scan the room. Slowly move your device around to scan the room, concentrating on the general area of the walls where you want to place the photos.

What makes for good plane detection?

When AR plane detection uses the device's built-in white light camera for scanning the 3D environment, it relies on good visual fidelity of the camera image. The room should be well lit. The surfaces being scanned should have distinctive and random textures to assist the detection software. For example, our AR Gallery project may have difficulty detecting vertical planes if your walls are too smooth. (Newer devices may include other sensors, such as laser-based **LIDAR** depth sensors that don't suffer from these limitations). If your device has trouble detecting vertical wall planes, try strategically adding some sticky notes or other markers on the walls to make the surfaces more distinctive to the software.

When at least one vertical plane is detected, the scan prompt will disappear, and you'll see the Main UI **Add** button. Tapping the **Add** button will enable **AddPicture Mode**, showing the **AddPicture** UI panel with its tap-to-place instructional graphic. When you tap a tracked plane, the **FramedPhoto** prefab will be instantiated in the scene. Here's what mine looks like, on the left side:

Figure 6.11 – Placing an object on the wall (left) and correcting for surface normal and upright (right)

Oops! The picture is sticking out of the wall perpendicularly, as shown in the preceding screenshot (on the left side). We want it to hang like a picture on the wall like in the right-hand image. Let's update the script to take care of this.

Completing the AddPictureMode script

There are a number of improvements we need to implement to complete the `AddPictureMode` script, including the following:

- Rotate the picture so it is upright and flat against the wall plane.
- Tell the picture which image to show in its frame.
- Include a default scale when the picture is first placed on a wall.

The `AddPictureMode` script contains the following line in the code that sets the rotation to `hitPose.rotation`:

```
Instantiate(placedPrefab, hitPose.position, hitPose.rotation);
```

As you can see in the previous screenshot, the "up" direction of a tracked plane is perpendicular to the surface of the plane, so with this code the picture appears to be sticking out of the wall. It makes sense to instantiate a placed object using this default up direction for horizontal planes, where you want your object standing up on the floor or a table. But in this project, we don't want to do that. We want the picture to be facing in the same direction as the wall. And we want it hanging straight up/down.

Instead of using the `hit.pose.rotation`, we should calculate the rotation using the plane's normal vector (`pose.up`). Then we call the `Quaternion.LookRotation` function to create a rotation with the specified forward and upward directions (see https://docs.unity3d.com/ScriptReference/Quaternion.LookRotation.html).

Quaternions

A quaternion is a mathematical construct that can be used to represent rotations in computer graphics. As a Unity developer, you simply need to know that rotations in Transforms use the `Quaternion` class. See https://docs.unity3d.com/ScriptReference/Quaternion.html. However, if you'd like an explanation of the underlying math, check out the great videos by *3Blue1Brown* such as *Quaternions and 3D rotation, explained interactively* at https://www.youtube.com/watch?v=zjMuIxRvygQ.

Another thing we need is the ability to tell the FramedPhoto which image to display. We'll add a public variable for the `imageInfo` that will be set by the Image Select menu (developed in the next section of this chapter).

Also, we will add a `defaultScale` property that scales the picture when it's instantiated. If you recall, we defined our prefab as normalized to 1 unit max size, which would make it 1 meter wide on the wall unless we scale it. We're only scaling the X and Y axes, leaving the Z at `1.0` so that the frame's depth is not scaled too. I'll set the default scale to `0.5`, but you can change it later in the Inspector.

Modify the `AddPictureMode` script as follows:

1. Add the following declarations at the top of the class:

```
public ImageInfo imageInfo;
[SerializeField] float defaultScale = 0.5f;
```

2. Replace the `PlaceObject` function with the following:

```
void PlaceObject(Vector2 touchPosition)
{
    if (raycaster.Raycast(touchPosition, hits,
        TrackableType.PlaneWithinPolygon))
    {
        ARRaycastHit hit = hits[0];

        Vector3 position = hit.pose.position;
        Vector3 normal = -hit.pose.up;
        Quaternion rotation = Quaternion.LookRotation
            (normal, Vector3.up);

        GameObject spawned = Instantiate(placedPrefab,
            position, rotation);

        FramedPhoto picture =
            spawned.GetComponent<FramedPhoto>();
        picture.SetImage(imageInfo);

        spawned.transform.localScale = new
            Vector3(defaultScale, defaultScale, 1.0f);
        InteractionController.EnableMode("Main");
```

```
            }
        }
```

3. Save the script and return to Unity.

Note that I had to negate the wall plane normal vector (`-hit.pose.up`), because when we created our prefab, by convention, the picture is facing in the minus-Z direction.

When you place a picture, it should now hang properly upright and be flush against the wall, as shown in right-hand panel of the screenshot at the top of this section.

Showing tracked planes in AddPicture mode

Another enhancement might be to hide the tracked planes while in Main mode and show them while in `AddPicture` mode. This would allow the user to enjoy their image gallery without that distraction. Take a look at how we did that in the *Hiding tracked object when not needed* topic of *Chapter 5, Using the AR User Framework*. At that time, we wrote a script, `ShowTrackablesOnEnable`, that we can use now too. Follow these steps:

1. With the **AddPicture Mode** game object selected in the **Hierarchy** (under **Interaction Controller**).

2. In the **Project** window, locate your `ShowTrackablesOnEnable` script and drag it onto the **AddPicture Mode** object.

3. From the **Hierarchy**, drag the **AR Session Origin** object into the **Inspector** and drop it onto the **Show Trackables On Enable | Session Origin** slot.

That is all we need to implement this feature.

To recap, we configured the scene to detect and track vertical planes, for the walls of your room. Then we created an **AddPicture UI** panel that prompts the user with an instructional graphic to tap to place. Next, we created an AddPicture mode, including the interaction **AddPicture Mode** game object and added a new `AddPictureMode` script. The script instantiates a copy of the **FramedPhoto** prefab when the user taps on a vertical plane. Then we improved the script by ensuring the picture is oriented flat on the wall and upright. The script also lets us change the image in the frame and its scale. Lastly, we display the trackable planes when in AddPicture mode and hide them when we return to Main mode.

The next step is to give the user a choice to select an image before hanging a new picture on the wall. We can now go ahead and create an image select menu for the user to pick one to use.

Selecting an image to use

The next thing we want to do is create an image select menu containing image buttons for the user to choose a photo before adding it to the scene. When the **Add** button is pressed, rather than immediately prompting the user to place a picture on the wall, we'll now present a menu of pictures to select from before hanging the image chosen on the wall. I'll call this *SelectImage mode*. We'll need to write an `ImageButtons` script that builds the menu using the **Images** list you've already added to the project (the **Image Data** game object). And then we'll insert the **SelectImage** mode before **AddPicture** mode, so the selected image is the one placed on the wall. Let's define the **SelectImage** mode first.

Creating the SelectImage mode

When `SelectImage` mode is enabled by the user, all we need to do is display the SelectImage UI menu panel with buttons for the user to pick which image to use. Clicking a button will notify the mode script by calling the public function, `SetSelectedImage`, that in turn tells the `AddPictureMode` which image to use.

Create a new C# script named `SelectImageMode` and write it as follows:

```
using UnityEngine;

public class SelectImageMode : MonoBehaviour
{
    void OnEnable()
    {
        UIController.ShowUI("SelectImage");
    }
}
```

Simple. When `SelectImageMode` is enabled, we display the **SelectImage UI** panel (containing the buttons menu).

Now we can add it to the **Interaction Controller** as follows:

1. In the **Hierarchy** window, *right-click* the **Interaction Controller** game object and select **Create Empty**. Rename the new object `SelectImage Mode`.

2. Drag the `SelectImageMode` script from the **Project** window onto the **SelectImage Mode** object adding it as a component.

3. Now we'll add the mode to the **Interaction Controller**. In the **Hierarchy**, select the **Interaction Controller** object.

4. In the **Inspector**, at the bottom-right of the **Interaction Controller** component, click the **+** button to add an item to the **Interaction Modes** dictionary.

5. Enter Select Image in the **Id** field.

6. Drag the **SelectImage Mode** game object from the **Hierarchy** onto the **Value** slot. The **Interaction Controller** component now looks like the following:

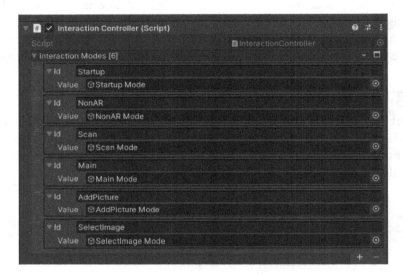

Figure 6.12 – Interaction Controller with SelectImage Mode added

Next, we'll add the UI for this mode.

Creating the Select Image UI panel

To create the **SelectImage UI** panel, we'll duplicate the existing **Main UI** and adapt it. The panel will include a Header title and **Cancel** button. Follow these steps:

1. In the **Hierarchy**, *right-click* the **Main UI** (child of **UI Canvas**) and select **Duplicate**. Rename the copy Select Image UI. Delete any child objects, including **Add Button**, using *right-click* **Delete**.

2. Make the panel size a little smaller than fullscreen so it looks like a modal popup. In its **Inspector** window, on the **Rect Transform**, set the **Left**, **Right**, and **Bottom** values to 50. Set the **Top** to 150 to leave room for the app title.

3. We want this panel to have a solid background so on its **Image** component, select **Component | UI | Image** from the main menu.

4. Create a menu header sub-panel. In the **Hierarchy**, *right-click* the **SelectImage** UI and select **UI | Panel** and rename it Header.

5. Position and stretch the **Header** to the top of the panel using the **Anchor Presets** box (icon in the upper-left of the **Rect Transform** component), click on **Top-Stretch**, and then *Alt+Shift* + click **Top-Stretch**. We can leave the default **Height** at 100.

6. *Right-click* the **Header**, select **UI | Text – TextMeshPro**, and rename the object Header Text.

7. On the **Header Text**, set its **Text** value to Select Image, **Vertex Color**: black, **Font Size** to 48, **Alignment** to **Center** and **Middle**, and **Anchor Presets** to **Stretch-Stretch**. Also, *Alt + Shift* + click **Stretch-Stretch**.

8. We'll also add a **Cancel** button to the Header. *Right-click* the **Header** object and select **UI | Button – TextMeshPro**. Rename it Cancel Button.

9. Set the Cancel button's **Anchor Preset** to **Middle-Right** and use *Alt + Shift* + click **Middle-Right** to position it there. Set its **Width, Height** to 80 and **Pos X** to -20. Also set its **Image | Color** to a light gray color.

10. For the **Cancel Button** child text element, set its **Text** value to X and **Font Size** to 48.

11. With the **Cancel Button** selected in **Hierarchy,** in its **Inspector** click the + button on the bottom-right of the **Button | OnClick** actions.

12. Drag the **Interaction Controller** game object from the **Hierarchy** onto the **OnClick Object** slot.

13. From the **Function** list, choose **InteractionController | EnableMode**. Enter Main in the text parameter field. The **X** cancel button will now send you back to Main mode.

The header of the **SelectImage** UI panel is shown in the following screenshot:

Figure 6.13 – The header panel of the SelectImage UI

Next, we'll add a panel to contain the image buttons that will display photos for the user to pick. These will be laid out in a grid. Use the following steps:

1. In the **Hierarchy,** *right-click* the **SelectImage UI**, select **UI | Panel**, and rename it Image Buttons.

2. On the **Image Buttons** panel, uncheck its **Image** component or remove it. We don't need a separate background.

3. Its **Anchor Presets** should already be **Stretch-Stretch** by default. Set the **Top** to `100`.

4. Select **Add Component**, search for `layout` and add a **Grid Layout Group** component.

5. On the **Grid Layout Group**, set its **Padding** to `20, 20, 20, 20`, set **Cell Size** to `200, 200`, and set **Spacing** to `20, 20`. Set **Child Alignment** to **Upper Center**.

We now have an **ImageSelect UI** panel with a header and a container for the image buttons. Parts of the current hierarchy are shown in the following screenshot:

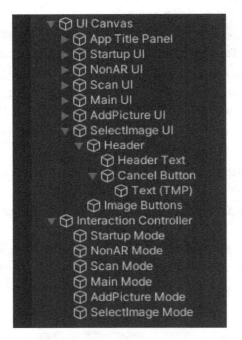

Figure 6.14 – UI Canvas with SelectImage UI, and Interaction Controller with SelectImage Mode

Lastly, we need to add the panel to the UI Controller as follows:

1. To add the panel to the UI Controller, in the **Hierarchy**, select the **UI Canvas** object.

2. In the **Inspector**, at the bottom-right of the **UI Controller** component, click the + button to add an item to the **UI Panels** dictionary.

3. Enter `SelectImage` in the **Id** field.

4. Drag the **SelectImage UI** game object from the **Hierarchy** onto the **Value** slot.

We now have a UI panel with a container for the image buttons. To make the buttons, we'll create a prefab and then write a script to populate the **Image Buttons** panel.

Creating an Image Button prefab

We will define an **Image Button** as a prefab so it can be duplicated for each image that we want to provide to the user in the selection menu. Create the button as follows:

1. In the **Hierarchy**, *right-click* the **Image Buttons** object, select **UI | Button**, and rename it `Image Button`.

2. Under the **Image Button**, delete its child **Text** element.

3. On the **Image Button**, remove its **Image** component (using *right-click* and **Remove Component**) and then press **Add Component**. Search and add a **Raw Image** component instead.

4. Its **Button** component needs a reference to its graphic that we just replaced. In the **Inspector**, drag the **Raw Image** component onto **Button | Target Graphic** slot.

5. Now drag your default image texture, such as the `WinterBarn` asset, from the **Project** window `Photos/` folder into the **Inspector** and drop it onto the **Raw Image | Texture** slot.

> **UI Image versus Raw Image**
>
> An **Image** component takes an image sprite for its graphic. A **Raw Image** component takes a texture for its graphic. Sprites are small, highly efficient, preprocessed images used for UI and 2D applications. Textures tend to be larger with more pixel depth and fidelity used for 3D rendering and photographic images. You can change an imported image between these and other type using the image file's Inspector properties. To use the same photo asset (PNG files) for both the FramedPhoto prefab and the button, we're using a Raw Image component on the buttons.

6. Let's save the **Image Button** as a prefab. Drag the **Image Button** object from the **Hierarchy** into the **Project** window and drop it into your Assets `Prefabs/` folder. This creates a prefab asset and changes its color in the **Hierarchy** to blue, indicating it's a prefab instance.

7. In the **Hierarchy** window, *right-click* the **Image Button** object, select **Duplicate** (or press *Ctrl/Option + D* on the keyboard), and make several copies. Because the buttons are parented by the **Image Buttons** panel that has a **Grid Layout Group**, they are rendered in a grid, as shown in the following screenshot:

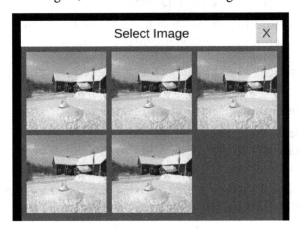

Figure 6.15 – Select image panel with Image Buttons in a grid layout

Next, we'll write a script to populate the buttons with actual images we want to use.

Writing an ImageButtons script

The ImageButtons script will be a component on the **Image Buttons** panel. Its job is to generate the image buttons with pictures of the corresponding images. Create a new C# script named ImageButtons, open it for editing, and write it as follows:

```
using UnityEngine;
using UnityEngine.UI;

public class ImageButtons : MonoBehaviour
{
    [SerializeField] GameObject imageButtonPrefab;
    [SerializeField] ImagesData imagesData;
    [SerializeField] AddPictureMode addPicture;

    void Start()
    {
        for (int i = transform.childCount - 1; i >= 0; i--)
        {
```

```
            GameObject.Destroy(
                transform.GetChild(i).gameObject);
        }

    foreach (ImageInfo image in imagesData.images)
    {
        GameObject obj =
            Instantiate(imageButtonPrefab,transform);
        RawImage rawimage = obj.GetComponent<RawImage>();
        rawimage.texture = image.texture;
        Button button = obj.GetComponent<Button>();
        button.onClick.AddListener(() => OnClick(image));
    }
    }

    void OnClick(ImageInfo image)
    {
        addPicture.imageInfo = image;
        InteractionController.EnableMode("AddPicture");
    }
}
```

Let's go through this script. At the top of the class, we declare three variables. `imageButtonPrefab` will be a reference to the **ButtonPrefab** that we will instantiated. `imagesData` is a reference to the object containing our list of images. And `addPicture` is a reference to `AddPictureMode` for each button to tell which image has been selected.

The first thing `Start()` does is clear out any child objects in the buttons panel. For example, we created a number of duplicates of the button to help us develop and visualize the panel, and they'll still be in the scene when it runs unless we remove them first.

Then, `Start` loops through each of the images, and for each one, creates an **Image Button** instance and assigns the image to the button's **RawImage** texture. And it adds a listener to the button's `onClick` events.

When one of the buttons is clicked, our `OnClick` function will be called, with that button's `image` as a parameter. We pass this `image` data to the `AddPictureMode` that will be used when **AddPictureMode** instantiates a new **FramedPhoto** object.

Add the script to the scene as follows:

1. In the **Hierarchy**, select the **Image Buttons** object (under **UI Canvas/Select Image Panel**).

2. Drag the **ImageButtons** script onto the **Image Buttons** object, making it a component.

3. From the **Project** window, drag the **Image Button** prefab into the **Inspector** and drop it onto the **Image Buttons | Image Button Prefab** slot.

4. From the **Hierarchy**, drag the **AddPicture Mode** object into the **Inspector** and drop it onto the **Image Buttons | Add Picture** slot.

5. Also from the **Hierarchy**, drag the **Images Data** object and drop it onto the **Image Buttons | Images Data** slot.

The **Image Buttons** component now looks like the following screenshot:

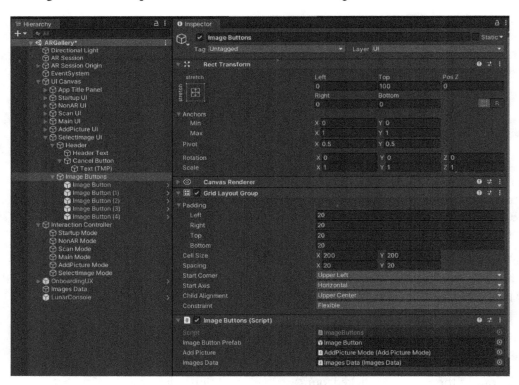

Figure 6.16 – Image buttons panel with the ImageButtons script that builds the menu at runtime

OK. When the app starts up, the **Image Buttons** menu will be populated from the **Images** list in **Images Data**. Then, when the user presses an image button, it'll tell the AddPictureMode which image was selected, and then enabled AddPicture mode.

Reroute the Add button

There is just one last step before we can try it out. Currently, the main menu's Add button enables AddPicture mode directly. We need to change it to call **SelectImage** instead, as follows:

1. In the **Hierarchy**, select the **Add** button (located in **UI Canvas / Main UI**).

2. In the **Inspector**, in the **Button | On Click** action list, change the **EnableMode** parameter from **AddPicture** to **SelectImage**.

3. Save the scene.

If you got all this right, you should be able to **Build and Run** the scene and run through the complete scenario: pressing the **Add** button will present a **Select Image** menu. Tapping an image, the select panel is replaced with a prompt to tap to place the image, with its frame, on a wall. The following screenshots from my phone show, on the left, the **Select Image** menu. After selecting an image and placing it on the wall, the result is shown on the right. Then the app returns to the main menu:

Figure 6.17 – Pressing the Add button, I see an image menu (left), and the result after placing (right)

To summarize, in this section we added the **Select Image** menu to the scene by first creating the UI panel and adding it to the UI Controller. Then we created an **Image Button** prefab and wrote the `ImageButtons` script that instantiates buttons for each image we want to include in the app. Clicking one of the buttons will pass the selected image data to **AddPicture** mode. When the user taps to place and a **FramedPhoto** is instantiated, we set the image to the one the user has selected. We also included a **Cancel** button in the menu so the user can cancel the add operation.

This is looking good so far. One problem we have is all the pictures are rendered in the same sized landscape frame and thus may look distorted. Let's fix that.

Adjusting for image aspect ratio

Currently, we're ignoring the actual size of the images and making them all fit into a landscape orientation with a 3:4 aspect ratio. Fortunately, we've included the actual (original) pixel dimensions of the image with our `ImageInfo`. We can use that now to scale the picture accordingly. We can make this change to the `FramedPhoto` script that's on the **FramedPhoto** prefab.

The algorithm for calculating the aspect ratio can be separated as a utility function in the `ImagesData` script. Open the `ImagesData` script and add the following code:

```
public static Vector2 AspectRatio(float width, float
    height)
{
    Vector2 scale = Vector2.one;

    if (width == 0 || height == 0)
        return scale;

    if (width > height)
    {
        scale.x = 1f;
        scale.y = height / width;
    }
    else
    {
        scale.x = width / height;
        scale.y = 1f;
    }
```

```
        return scale;
    }
```

When the `width` is larger than `height`, the image is landscape, so we'll keep the X scale at `1.0` and scale down Y. When the `height` is larger than the `width`, it is portrait, so we'll keep the Y scale at `1.0` and scale down X. If they're the same or zero, we return `(1,1)`. The function is declared `static` so it can be called using the `ImagesData` class name.

Open the `FramedPhoto` script for editing and make the changes highlighted in the following:

```
    public void SetImage(ImageData image)
    {
        imageData = image;

        Renderer renderer =
            imageObject.GetComponent<Renderer>();
        Material material = renderer.material;
        material.SetTexture("_BaseMap", imageData.texture);
        AdjustScale();
    }

    public void AdjustScale()
    {
        Vector2 scale = ImagesData.AspectRatio(imageInfo.width,
            imageInfo.height);
        scalerObject.localScale = new Vector3(scale.x, scale.y,
            1f);
    }
```

If you recall, the `SetImage` function is called by `AddPictureMode` immediately after a **FramedPhoto** object is instantiated. After **SetImage** sets the texture, it now calls **AdjustScale** to correct its aspect ratio. **AdjustScale** uses `ImageData.AspectRatio` to get the new local scale and updates the `scalerObject` transform.

You may notice that the frame width is slightly different on the horizontal versus vertical sides when the picture is not square. Fixing this requires an additional adjustment to the **Frame** object's scale. For example, on a landscape orientation, try setting the child **Frame** object's **Scale X** to `1.0 - 0.01/aspectratio`. I'll leave that implementation up to you.

When you run the project again and place a picture on your wall, it'll be the correct aspect ratio according to the photo you picked. One improvement you could add is to scale the images on the **Select Image Panel** buttons so they too are not squished. I'll leave that exercise up to you.

Summary

At the beginning of this chapter, I gave you the requirements and a plan for this AR gallery project, including a statement of the project objectives, use cases, UX design, and user stories. You started the implementation using the **ARFramework** template created in *Chapter 4, Creating an AR User Framework*, and built upon it to implement new features for placing a framed photo on your walls.

To implement this feature, you created a **SelectImage UI** panel, a **SelectImage Mode** interaction mode, and populated a list of images data. After the app starts up and AR is tracking vertical planes, when the user presses the **Add** button in the main menu, it opens a **Select Image** menu showing images to pick from. The image buttons grid was generated from your image data using an `ImageButton` prefab you created. Clicking an image, you're prompted to tap an AR tracked wall, and a new framed photo of that image is placed on the wall, correctly scaled to the image's aspect ratio.

We now have a fine start to an interesting project. There is a lot more that can be done. For example, presently pictures can be placed on top of one another, which would be a mistake. Also, it would be good to be able to move, resize, and remove pictures. We'll add that functionality in the next chapter.

7
Gallery: Editing Virtual Objects

In this chapter, we will continue building the project we started previously in *Chapter 6, Gallery: Building an AR App*, where we created an AR gallery that lets users place virtual framed photos on their real-world walls. In this chapter, we will build out more features related to interacting with and editing virtual objects that have already been added to a scene. Specifically, we'll let users select an object for editing, including moving, resizing, deleting, and replacing the image in the picture frame. In the process, we'll add new input actions, make use of Unity collision detection, and see more C# coding techniques using the Unity API.

In this chapter, we will cover the following topics:

- Detecting collisions to avoid intersecting objects
- Building an edit mode and edit menu UI
- Using a physics raycast to select an object
- Adding touch input actions to drag to move and pinch to scale
- C# coding and the Unity API, including collision hooks and vector geometry

By the end of this chapter, you'll have a working AR application with many user interactions implemented.

Technical requirements

To complete the project in this chapter, you will need Unity installed on your development computer, connected to a mobile device that supports augmented reality applications (see *Chapter 1, Setting Up for AR Development*, for instructions). We will also assume you have created the *ARGallery* scene that we started in *Chapter 6, Gallery: Building an AR App*, where you'll also find additional dependencies detailed for you in the *Technical requirements* section. You can find that scene, as well as the one we will build in this chapter, in this book's GitHub repository at `https://github.com/PacktPublishing/Augmented-Reality-with-Unity-AR-Foundation`.

Note that in this book's repository, some of the scripts (and classes) for this chapter have been post-fixed with `07`, such as `AddPictureMode07`, to distinguish them from the corresponding scripts that were written for the previous chapter. In your own project, you can leave the un-post-fixed name as is when you edit the existing scripts described in this chapter.

Creating an Edit mode

To get started with this chapter, you should have the *ARGallery* scene open in Unity where we left off at the end of *Chapter 6, Gallery: Building an AR App*. To recap, after launching the app, it starts by initializing the AR session and scanning to detect features in your real-world environment. Once the vertical planes (walls) have been detected, the main menu will be presented. Here, the user can tap the **Add** button, which opens an image select menu where the user can pick a photo to use. Then, the user will be prompted to tap on a trackable vertical plane to place the framed photo on. Once the photo is hanging on their wall, the user is returned to Main-mode.

In this chapter, we'll let users modify existing virtual framed photos that have been added to the scene. The first step is for the user to select an existing object to edit from Main-mode, which then activates EditPicture-mode for the selected object. When an object is selected and being edited, it should be highlighted so that it's apparent which object has been selected.

Using the AR user framework that's been developed for this book, we will start by adding an EditPicture-mode UI to the scene. First, we'll create the edit menu user interface, including multiple buttons for various edit functions, and an Edit-mode controller script for managing it.

Creating an edit menu UI

To create the UI for editing a placed picture, we'll make a new **EditPicture UI** panel. It's simpler to duplicate the existing **Main UI** and adapt it. Perform the following steps:

1. In the **Hierarchy** window, *right-click* **Main UI** (child of **UI Canvas**) and select **Duplicate**. Rename the copy EditPicture UI. Delete any child objects, including **Add Button**, by *right-clicking* | **Delete**.

2. Create a subpanel for the menu by *right-clicking* **EditPicture UI** and selecting **UI** | **Panel**. Rename it Edit Menu.

3. Use **the Anchors presets** to set **Bottom-Stretch**, and then use *Shift* + *Alt* + **Bottom-Stretch** to make a bottom panel. Then, set its **Rect Transform** | **Height** value to 175.

4. I set my background **Image** | **Color** to opaque white with **Alpha at**55.

5. Select **Add Component**, search for layout, and select **Horizontal Layout Group**.

6. On the **Horizontal Layout Group** component, check the **Control Child Size** | **Width** and **Height** checkboxes. (Leave the others at their default values, **Use Child Scale** unchecked, and **Child Force Expand** checked). The **Edit Menu** panel looks like this in the **Inspector** window:

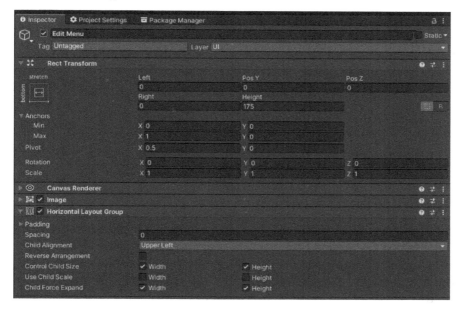

Figure 7.1 – The Edit Menu panel settings

7. Now, we will add four buttons to the menu. Begin by *right-clicking* **Edit Menu** and selecting **UI | Button – TextMeshPro**. Rename it `Replace Image Button`.

8. Select its child text object, set the **Text** value to `Replace Image`, and set **Font Size** to `48`.

9. *Right-click* the **Replace Image** button and select **Duplicate** (or *Ctrl + D*). Repeat this two more times so that there are four buttons in total.

10. Rename the buttons and change the text on the buttons so that they read as `Replace Frame`, `Remove Picture`, and `Done`.

11. We may not use the **Replace Frame** feature soon, so disable that button by unchecking its **Interactable** checkbox in the **Button** component. The resulting menu will look as follows:

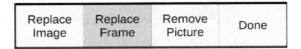

Figure 7.2 – Edit Menu buttons

Add the panel to the UI Controller, as follows:

1. To add the panel to the UI Controller, in the **Hierarchy** window, select the **UI Canvas** object.

2. In the **Inspector** window, at the bottom right of the **UI Controller** component, click the + button to add an item to the UI Panels dictionary.

3. Enter `EditPicture` in the **Id** field.

4. Drag the **EditPicture UI** game object from the **Hierarchy** window onto the **Value** slot.

The next step is to create an **EditPicture** mode object and controller script.

Creating EditPicture mode

As you now know, our framework manages interaction modes by activating game objects under the Interaction Controller. Each mode has a control script that displays the UI for that mode and handles any user interactions until certain conditions are met; then, it transitions to a different mode. In terms of our EditPicture-mode, its control script will have a `currentPicture` variable that specifies which picture is being edited, a `DoneEditing` function that returns the user to Main-mode, among other features.

Create a new C# script named `EditPictureMode` and begin to write it, as follows:

```
using UnityEngine;

public class EditPictureMode : MonoBehaviour
{
    public FramedPhoto currentPicture;

    void OnEnable()
    {
        UIController.ShowUI("EditPicture");
    }
}
```

Now, we can add it to our **Interaction Controller** object, as follows:

1. In the **Hierarchy** window, *right-click* the **Interaction Controller** game object and select **Create Empty**. Rename the new object `EditPicture Mode`.

2. Drag the `EditPictureMode` script from the **Project** window onto the **EditPicture Mode** object, adding it as a component.

3. Now, we'll add the mode to the Interaction Controller. In the **Hierarchy** window, select the **Interaction Controller** object.

4. In the **Inspector** window, at the bottom right of the **Interaction Controller** component, click the + button to add an item to the **Interaction Modes** dictionary.

5. Enter `EditPicture` in the **Id** field.

6. Drag the **EditPicture Mode** game object from the **Hierarchy** window onto the **Value** slot.

With that, we have created an **EditPicture UI** containing edit buttons that is controlled by `UIController`. After this, we created an **EditPicture Mode** game object with an `EditPictureMode` script that is controlled by `InteractionController`.

With this set up, the next thing we must do is enhance Main-mode so that it detects when the user taps on an existing **FramedPhoto** and can start EditPicture-mode for the selected object.

Selecting a picture to edit

While in Main-mode, the user should be able to tap on an existing picture to edit it. Utilizing the Unity Input System, we will add a new `SelectObject` input action. Then, we'll have the `MainMode` script listen for that action's messages, find which picture was tapped using a `Raycast`, and enable Edit-mode on that picture. Let's get started!

Defining a SelectObject input action

We will start by adding a `SelectObject` action to the **AR Input Actions** asset by performing the following steps:

1. In the **Project** window, locate and *double-click* the **AR Input Actions** asset we created previously (it may be in the `Assets/Inputs/` folder) to open it for editing (alternatively, use its **Edit Asset** button).

2. In the middle **Actions** section, select + and name it `SelectObject`.

3. In the rightmost **Properties** section, select **Action Type | Value** and **Control Type | Vector 2**.

4. In the middle **Actions** section, select the **<No Binding>** child. Then, in the **Properties** section, select **Path | Touchscreen | Primary Touch | Position** to bind this action to a primary screen touch.

5. Press **Save Asset** (unless **Auto-Save** is enabled).

The updated **AR Input Actions** asset is shown in the following screenshot:

Figure 7.3 – AR Input Actions asset with the SelectObject action added

Although we're defining this action with the same touchscreen binding that we used for the `PlaceObject` action we created earlier (**Touchscreen Primary Position**), it serves a somewhat different purpose (tap-to-select versus tap-to-place). For example, perhaps, in the future, if you decide to use a *double-tap* for selecting an item instead of a single tap, you can simply change its input action.

Now, we can add the code for this action.

Replacing the MainMode script

First, because we're deviating from the default `MainMode` script provided in the `ARFramework` template, we should make a new, separate script for this project. Perform the following steps to copy and edit the new `GalleryMainMode` script:

1. In the `Project` window's `Scripts/` folder, select the **MainMode** script. Then, from the main menu bar, select **Edit | Duplicate**.

2. Rename the new file `GalleryMainMode`.

3. You'll see a namespace error in the **Console** window because we now have two files defining the `MainMode` class.

   Open **GalleryMainMode** for editing and change the class name to `GalleryMainMode`, as highlighted here:

   ```
   using UnityEngine;

   using UnityEngine.InputSystem;

   public class GalleryMainMode : MonoBehaviour
   {
       void OnEnable()
       {
           UIController.ShowUI("Main");
       }
   }
   ```

4. Save the script. Then, back in Unity, in the **Hierarchy** window, select the **Main Mode** game object (under **Interaction Controller**).

5. Drag the **GalleryMainMode** script onto the **Main Mode** object, adding it as a new component.

6. Remove the previous **Main Mode** component from the **Main Mode** object.

Now, we're ready to enhance the behavior of Main-mode.

Selecting an object from Main-mode

When the user taps the screen, the `GalleryMainMode` script will get the touch position and use a Raycast to determine whether one of the `PlacedPhoto` objects was selected. If so, it will enable **EditPicture** mode on that picture.

We have seen Raycasts previously in our tap-to-place scripts, including `AddPictureMode`. In that case, our scripts used the **AR Raycast Manager** class's version of the function because we were only interested in hitting a tracked AR plane. But in this case, we're interested in selecting a regular GameObject – an instantiated FramedPhoto prefab. For this, we'll use the `Physics.Raycast` function (`https://docs.unity3d.com/ScriptReference/Physics.Raycast.html`). As part of the Unity Physics system, it requires the raycast-able object to have a **Collider** (which **FramedPhoto** does, and I'll show you soon).

Also, we will be using the AR Camera's `ScreenPointToRay` function to define the 3D ray that corresponds to the touch position that we're going to Raycast into the scene.

To add this, open the `GalleryMainMode` script for editing and follow these steps:

1. We're going to listen for Input System events, so to begin, we need to add a `using` statement for that namespace. Ensure the following line is at the top of the file:

    ```
    using UnityEngine.InputSystem;
    ```

2. We need a reference to tell `EditPictureMode` which object to edit. Add it to the top of the class, as follows:

    ```
    public class GalleryMainMode : MonoBehaviour
    {
        [SerializeField] EditPictureMode editMode;
    ```

3. We're going to be using **AR Camera** here, so it's good practice to get a reference to that at the start by using the `Camera.main` shortcut. (This requires the AR Camera to be tagged as `MainCamera`, which should be done from the scene template.) Add a private variable at the top of the class and initialize it using `Start`:

    ```
    Camera camera;

    void Start()
    {
        camera = Camera.main;
    }
    ```

4. Now for the meat of our task – add the following `OnSelectObject` and `FindObjectToEdit` functions:

```
public void OnSelectObject(InputValue value)
{
    Vector2 touchPosition = value.Get<Vector2>();
    FindObjectToEdit(touchPosition);
}


void FindObjectToEdit(Vector2 touchPosition)
{
    Ray ray = camera.ScreenPointToRay(touchPosition);
    RaycastHit hit;
    int layerMask =
        1 << LayerMask.NameToLayer("PlacedObjects");
    if (Physics.Raycast(ray, out hit, Mathf.Infinity,
        layerMask))
    {
        FramedPhoto picture = hit.collider.
            GetComponentInParent<FramedPhoto>();
        editMode.currentPicture = picture;
        InteractionController.
            EnableMode("EditPicture");
    }
}
```

Let's walk through this code together. The `OnSelectObject` function is automatically called when the `SelectObject` Input System action is used (the On prefix is a standard Unity convention for event interfaces). It grabs `Vector2 touchPosition` from the input value and passes it to our private `FindObjectToEdit` function. (You don't need to separate this into two functions, but I did for clarity.)

`FindObjectToEdit` gets the 3D ray corresponding to the touch position by calling `camera.ScreenPointToRay`. This is passed to `Physics.Raycast` to find an object in the scene that intersects with the ray. Rather than casting to every possible object, we'll limit it to ones on a layer named `PlacedObjects` using its `layermask`. (For this, we need to make sure **FramedPhoto** is assigned to this layer, which we'll do soon.)

> **Information – Layer Name, Layer Number, and Layermask**
>
> A **layermask** uses the binary bits of a 32-bit integer to identify up to 32 layers, one bit each. We define the mask by getting the layer number from its name (`LayerMask.NameToLayer`) and shifting one bit to the left that many times. To manage the layers in your project and see what name has been assigned to each layer number, click the **Layers** button in the top-right corner of the Editor.

If the raycast gets a hit, we must grab a reference to the `FramedPhoto` component in the prefab and pass it to the `EditPictureMode` component. Then, the app will transition to EditPicture-mode.

Save the script. Now, let's fix the housekeeping things on our game objects that I mentioned: set the camera tag to `MainCamera`, set the **FramedPhoto** object so that it resides on the `PlacedObjects` layer, and ensure **FramedPhoto** has a collider component. In Unity, do the following:

1. In the **Hierarchy** window, with the **Main Mode** game object selected, drag the **EditPicture Mode** object from the **Hierarchy** window into the **Inspector** window and drop it onto the **Gallery Main Mode | Edit Mode** slot.

2. In the scene **Hierarchy**, unfold **AR Session Origin** and select its child **AR Camera**. In the top-left position of the **Inspector** window, verify that **Tag** (atop the **Inspector** window) is set to **MainCamera**. If not, set it now.

3. Next, open the **FramedPhoto** prefab for editing by *double-clicking* the asset in the **Project** window.

4. With its root **FramedPhoto** object selected, in the top right of its **Inspector** window, click the **Layer** drop-down list and select `PlacedObjects`.

   If the layer named `PlacedObjects` doesn't exist, select **Add Layer** to open the **Layers manager** window. Add `PlacedObjects` to one of the empty slots. In the **Hierarchy** window, click the **FramedPhoto Prefab** object to get back to its **Inspector** window. Again, using the **Layers** drop-down list, select `PlacedObjects`.

   You will then be prompted with the question **Do you want to set layer to PlacedObjects for all child objects as well?**. Click **Yes, Change Children**.

5. While we're here, let's also verify that the prefab has a collider, as required for `Physics.Raycast`. If you recall, when we constructed the prefab, we started with an **Empty** game object for the root and added another **Empty** child for **AspectScaler**. Then, we added a 3D Cube for the **Frame** object. Click this **Frame** object.

6. In the **Inspector** window, you will see that the **Frame** object already has a **Box Collider**. Perfect. Note that if you press its **Edit Collider** button, you can see (and edit) the collider's shape in the **Scene** window, as shown in the following screenshot, where its edges are outlined and there are little handles to move the faces. But there's no need for us to change it here:

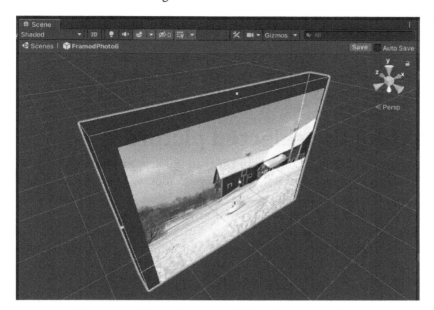

Figure 7.4 – Editing the Box Collider of the Frame object

7. Save the prefab and exit the prefab editor to get back to the Scene hierarchy.

If you were to **Build and Run** the scene now, and then add a picture to a wall, when you tap on that picture, it should hide the main menu and show the edit menu. Now, we need a way to get back from Edit-mode to Main-mode. Let's wire up the **Done** button.

Wiring the Done edit button

In this section, we will set up the **Done** button to switch from EditPicture-mode to Main-mode. It simply needs to call `EnableMode` in `InteractionController`. Follow these steps:

1. In the **Hierarchy** window, select the **Done** button, which should be located under **UI Canvas | EditPicture UI | Edit Menu**.

2. In the **Inspector** window, click the + button on the bottom right of the **Button | OnClick** area to add a new event action.

3. Drag the **Interaction Controller** object from the **Hierarchy** window and drop it onto the **Object** slot of the **OnClick** action.

4. In the function select list, choose **InteractionController | EnableMode**.

5. Type `Main` into the mode string parameter slot.

Now, if you **Build and Run** the scene where you have a picture instantiated in the scene and tap the picture, you'll switch to Edit-mode and see the edit menu. Tap the **Done** button to get back to Main-mode.

This is progress. But if there's more than one picture on your wall, it's not obvious which one is currently being edited. We need to highlight the currently selected picture.

Highlighting the selected picture

There are many ways to highlight objects in Unity to indicate that an object has been selected by the user. Often, you'll find that a custom shader will do the trick (there are many on the Asset Store). The decision comes down to what "look" you want. Do you want to change the selected object's color tint, draw a wireframe outline, or create some other effect? Instead of doing this and to keep things easy, I'll just introduce a "highlight" game object in the **FramedPhoto** prefab as a thin yellow box that extends from the edges of the frame. Let's make that now:

1. Open the **FramedPhoto** prefab for editing by *double-clicking* it in the **Project** window.

2. In the **Hierarchy** window, *right-click* on the **AspectScaler** object and select **3D Object | Cube**. Rename the cube `Highlight`.

3. Set its **Transform | Scale** setting to (`1.05, 1.05, 0.005`) so that it is thin and extends past the edges of the frame.

4. Set its **Transform | Position** setting to (`0, 0, -0.025`).

5. Create a yellow material. In the **Project** window, *right-click* in your `Materials/` folder (create one if needed) and select **Create | Material**. Rename it `Highlight Material`.

6. Set **Highlight Material | Shader | Universal Render Pipeline | Unlit**.

7. Set its **Base Map** color (using the color swatch) to yellow.

8. Drag **Highlight Material** onto the **Highlight** game object. The **Scene** view should now look as follows:

Figure 7.5 – FramedPhoto with highlight

We can now control this from the `FramedPhoto` script. You may want the highlight the picture for different reasons, but for this project, I've decided that when the object is selected and highlighted, that means it is being edited. So, we can toggle the highlight when making the object editable. Open the script in your editor and make the following changes:

1. Declare a variable for `highlightObject`:

    ```
    [SerializeField] GameObject highlightObject;
    bool isEditing;
    ```

2. Add a function to toggle the highlight:

    ```
    public void Highlight(bool show)
    {
        if (highlightObject) // handle no object or app
                             end object destroyed
            highlightObject.SetActive(show);
    }
    ```

3. Ensure the picture isn't highlighted at the beginning:

```
void Awake()
{
    Highlight(false);
}
```

4. Add a `BeingEdited` function. This will be called when the object is being edited. It'll highlight the object and enable other editing behavior later. Likewise, when we stop editing and pass a `false` value, the object will be un-highlighted:

```
public void BeingEdited(bool editing)
{
    Highlight(editing);
    isEditing = editing;
}
```

5. Save the script. In Unity, select the root **FramedPhoto** object.

6. Drag the **Highlight** object from the **Hierarchy** window onto the **Framed Photo | Highlight Object** slot.

This is great! Now, we can update `EditPictureMode` to tell the picture when it's being edited or not. Open the `EditPictureMode` script and make the following edits:

1. Add the `BeingEdited` call to `OnEnable`:

```
void OnEnable()
{
    UIController.ShowUI("EditPicture");
    if (currentPicture)
        currentPicture.BeingEdited(true);
}
```

2. Also, add the `BeingEdited` call to `OnDisable` for when it's not being edited; that is, when Edit-mode has been exited:

```
void OnDisable()
{
    if (currentPicture)
        currentPicture.BeingEdited(false);
}
```

Notice that although we would never intentionally enter Edit-mode without `currentPicture` defined, I've added null checks in case the mode is activated or deactivated during the app startup or teardown sequences.

If you play the scene now and add a picture, when you tap the picture via Main-mode, Edit-mode will become enabled, and the picture will be highlighted. When you exit back to Main-mode, the picture will be un-highlighted.

Let's keep going. Suppose you have multiple pictures on your walls. Currently, when you're editing one picture and you want to edit a different one, you must press **Done** to exit Edit-mode and then select the other picture from Main-mode. To switch between objects that are currently being editing, we can add that code to the `EditMode` script.

Selecting an object from Edit mode

When in Edit-mode for one picture, to let the user choose a different picture without exiting Edit-mode, we can use the same **SelectObject** input action we used in Main-mode. In fact, the code is mostly the same. Open the `EditPictureMode` script for editing and make the following changes:

1. We're going to listen for Input System events, so to begin, we need to add a `using` statement for that namespace. Ensure the following line is at the top of the file:

    ```
    using UnityEngine.InputSystem;
    ```

2. Add a private `camera` variable at the top of the class and initialize it in `Start`:

    ```
    Camera camera;


    void Start()
    {
        camera = Camera.main;
    }
    ```

3. The `OnSelectObject` action listener will call `FindObjectToEdit`. Like in `GalleryMainMode`, it does a Raycast on the `PlacedObjects` layer. But now, we must check whether it has hit an object other than the current picture. If so, we must stop editing `currentPicture` and make the new selection current:

    ```
    public void OnSelectObject(InputValue value)
    {
        Vector2 touchPosition = value.Get<Vector2>();
    ```

```
            FindObjectToEdit(touchPosition);
    }


    void FindObjectToEdit(Vector2 touchPosition)
    {
        Ray ray = camera.ScreenPointToRay(touchPosition);
        RaycastHit hit;
        int layerMask =
            1 << LayerMask.NameToLayer("PlacedObjects");
        if (Physics.Raycast(ray, out hit, 50f,
            layerMask))
        {
            if (hit.collider.gameObject !=
                currentPicture.gameObject)
            {
                currentPicture.BeingEdited(false);
                FramedPhoto picture = hit.collider.
                    GetComponentInParent<FramedPhoto>();
                currentPicture = picture;
                picture.BeingEdited(true);
            }
        }
    }
```

To summarize, when you have more than one **FramedPhoto** instantiated in the scene and you are editing one, if you tap on a different picture, the current one will be un-highlighted and the new one will be highlighted and become the `currentPicture` object being edited.

Here's another problem: if you've been playing with the project, you may have noticed that you can place pictures on top of one another, or actually, *inside* one another, as they do not seem to have any physical presence! Oops. Let's fix this.

Avoiding intersecting objects

In Unity, to specify that an object should participate in the Unity Physics system, you must add a **Rigidbody** component to the GameObject. Adding a Rigidbody gives an object mass, velocity, collision detection, and other physical properties. We can use this to prevent objects from intersecting. In many games and XR apps, Rigidbody is important for applying motion forces to objects to let them bounce when they collide, for example.

In our project, if a picture collides with another picture, it should simply move out of the way so that they're never intersecting. But it should also stay flush with the wall plane. Although a Rigidbody allows you to constrain movement along any of the **X**, **Y**, and **Z** directions, these are the orthogonal world space planes, not the arbitrary angled wall plane. In the end, I decided to position the picture manually when a collision is detected rather than using physics forces. My solution is to constrain the position (and rotation) of all the pictures so that physics forces won't move them. Then, I can use the collision as a trigger to manually move the picture out of the way.

Information – Collision Versus Trigger Detection

When two GameObjects with **Rigidbody** and **Collider** collide, physics forces will be applied to the objects, sending them in different directions. You can add constraints and other properties to limit this behavior. In that case, you can write functions for `OnCollisionEnter`, `OnCollisionStay`, and `OnCollisionExit` to hook into these events.

However, you can completely disable Unity applying physical forces by marking a Collider as **Is Trigger**. When it's a trigger, you would instead write functions for `OnTriggerEnter`, `OnTriggerStay`, and `OnTriggerExit` to hook into these events.

To add collision detection to the **FramedPhoto** prefab, follow these steps:

1. In the **Project** window, locate and *double-click* on the **FramedPhoto** prefab to open it for editing.

2. Ensure you have selected the root **FramedPhoto** object in the **Hierarchy** window.

3. In the **Inspector** window, click **Add Component**, search for `rigidbody`, and add a **Rigidbody** to the object.

4. Unfold the **Constraints** properties and check all six boxes; that is, **Freeze Position: X, Y, Z** and **Freeze Rotation: X, Y, Z**.

5. Uncheck its **Use Gravity** checkbox. (This is not necessary since we set constraints, but I like to be clear about this anyway.)

6. We need a **Collider**. As we've seen, there is one on the **Frame** child object. So, select the **Frame** game object.

7. In the **Inspector** window, in its **Box Collider** component, check the **Is Trigger** checkbox.

8. To avoid any problems, disable (or remove) other colliders in the prefab. Namely, remove **Mesh Collider** from **Image** and **Box Collider** from **Highlight**.

Now, we can handle the collision trigger and move the picture out of the way when another picture is in the same space. We just want to make sure it moves along the wall. We can make use of the fact that the wall plane's normal vector (the vector that's perpendicular to the surface of the plane) is also the forward direction vector of our picture prefab since we originally placed it there. Also, we only want to consider collisions with objects on the placed object plane (for example, not the AR tracked plane objects).

My algorithm determines the distance between this picture and the other intersecting picture, in 3D. Then, it finds the direction to move this picture in by projecting the distance vector onto the wall plane and scaling it. The picture will continue moving away from the other frames until it is no longer intersecting.

Let's write the code for this. Open the `FramedPhoto` script for editing and follow these steps:

1. Begin by adding a reference to the `collider` and `layer` numbers at the top of the class, as follows:

    ```
    [SerializeField] Collider boundingCollider;
    int layer;
    ```

2. Initialize the `layer` number from its name. It's good to initialize this ahead of time because `OnTriggerStay` may be called every frame:

    ```
    void Awake()
    {
        layer = LayerMask.NameToLayer("PlacedObjects");
        Highlight(false);
    }
    ```

3. We'll use `OnTriggerStay` here, which is called with each update while the object is colliding with another object, as follows:

    ```
    void OnTriggerStay(Collider other)
    {
        const float spacing = 0.1f;

        if (isEditing && other.gameObject.layer == layer)
    ```

```
        {
                Bounds bounds = boundingCollider.bounds;
                if (other.bounds.Intersects(bounds))
                {
                        Vector3 centerDistance =
                                bounds.center - other.bounds.center;
                        Vector3 distOnPlane =
                                Vector3.ProjectOnPlane(centerDistance,
                                        transform.forward);
                        Vector3 direction =
                                distOnPlane.normalized;
                        float distanceToMoveThisFrame =
                                bounds.size.x * spacing;
                        transform.Translate(direction *
                                distanceToMoveThisFrame);
                }
        }
}
```

4. Save the script. In Unity, drag the **Frame** object (which has a Box Collider) from the **Hierarchy** window onto the **Framed Photo | Bounding Collider** slot. The **Framed Photo** component now looks as follows:

Figure 7.6 – Framed Photo component properties, including Bounding Collider

5. Save the prefab and return to the scene **hierarchy**.

When you play the scene now, place a picture on a wall, and then place another picture in the same space, the new picture will move away from the first one until they're no longer colliding.

Now that we can have many pictures on our walls, you might want to learn how to remove one from the scene. We'll look at this in the next section.

Deleting a picture

Deleting the picture that is being edited is straightforward. We just need to destroy the `currentPicture` GameObject and go back to Main-mode. Perform the following steps:

1. Open the `EditPictureMode` script and add the following function:

    ```
    public void DeletePicture()
    {
        GameObject.Destroy(currentPicture.gameObject);
        InteractionController.EnableMode("Main");
    }
    ```

2. Save the script.
3. In Unity, in the **Hierarchy** window, select **Remove Button** (located under **UI Canvas | EditPicture UI | Edit Menu**).
4. In the **Inspector**, click the + button at the bottom right of the **Button | OnClick** area.
5. Drag the **EditPicture Mode** object from the **Hierarchy** window onto the **OnClick Object** slot.
6. From the function selection, choose **EditPictureMode | DeletePicture**.

When you play the scene, create a picture, go into EditPicture-mode, and then tap the **Remove Picture** button, the picture will be deleted from the scene, and you will be back in Main-mode.

We now have two of the Edit menu buttons operating – **Remove Picture** and **Done**. Now, let's add the feature that lets you change the picture in an existing **FramedPhoto** from the **Image Select** menu panel.

Replacing the picture's image

When you add a picture from the Main menu, the Select Image menu is displayed. From here, you can pick a picture. At this point, you will be prompted to add a **FramedPhoto** to the scene using the image you selected. We implemented this by adding a separate **SelectImage Mode**. We now want to make that mode serve two purposes. It's called from Main-mode when you're adding a new, framed photo to the scene, and it's called from EditPicture-mode when you want to replace the image of an existing framed photo that's already in the scene. This requires us to refactor the code.

Currently, when we build the Select Image buttons (in the `ImageButtons` script) we have it configure and enable AddPicture-mode directly. Instead, it now needs to depend on how SelectImage-mode is being used, so we'll move that code from `ImageButtons` to `SelectImageMode`, as follows:

1. Edit the `SelectImageMode` script and add a reference to `AddPictureMode` at the top of the class:

    ```
    [SerializeField] AddPictureMode addPicture;
    ```

2. Then, add a public `ImageSelected` function:

    ```
    public void ImageSelected(ImageInfo image)
    {
        addPicture.imageInfo = image;
        InteractionController.EnableMode("AddPicture");
    }
    ```

3. Edit the `ImageButtons` script and add a reference to `SelectImageMode` at the top of the class:

    ```
    [SerializeField] SelectImageMode selectImage;
    ```

4. Then, replace the `OnClick` code with a call to `ImageSelected`, which we just wrote:

    ```
    void OnClick(ImageInfo image)
    {
        selectImage.ImageSelected(image);
    }
    ```

This refactoring has not added any new functionality, but it restructures the code for `SelectImageMode` to decide how the modal menu will be used. Now, let's edit `SelectImageMode` again and add support for replacing the `currentPicture` image.

5. At the top of the `SelectImageMode` script, add the following declarations:

    ```
    [SerializeField] EditPictureMode editPicture;
    public bool isReplacing = false;
    ```

6.   Then, update the `ImageSelected` function, as follows:

```
public void ImageSelected(ImageInfo image)
{
    if (isReplacing)
    {
        editPicture.currentPicture.SetImage(image);
        InteractionController.
            EnableMode("EditPicture");
    }
    else
    {
        addPicture.imageInfo = image;
        InteractionController.
            EnableMode("AddPicture");
    }
}
```

So, now, when the menu is being used for replacing, it sends the selected image data to the edit mode's `currentPicture` object. Otherwise, it behaves as it did previously for AddPicture-mode.

Now, we need to make sure the `isReplacing` flag is set to `false` when *adding* and set to `true` when *replacing*. Again, this requires some refactoring. Currently, the main menu's **Add** button enables SelectImage-mode directly. Let's replace this with a `SelectImageToAdd` function in the `GalleryMainMode` script.

7.   At the top of the `GalleryMainMode` class, add a reference to `SelectImageMode`:

```
[SerializeField] SelectImageMode selectImage;
```

8.   Then, add a `SelectImageToAdd` function, as follows:

```
public void SelectImageToAdd ()
{
    selectImage.isReplacing = false;
    InteractionController.EnableMode("AddPicture");
}
```

We just need to remember to update the **Add** button **OnClick** action before we're done.

9. Likewise, now, we can add a `SelectImageToReplace` function to the `EditPictureMode` script. Declare `selectImage` at the top of the class:

```
[SerializeField] SelectImageMode selectImage;
```

Then, add the function, as follows:

```
public void SelectImageToReplace()
{
    selectImage.isReplacing = true;
    InteractionController.EnableMode("SelectImage");
}
```

Save all the scripts. Now, we need to connect it up in Unity, including setting the **Add** and **Replace Image** buttons' **OnClick** actions, and then setting the new **SelectImage Mode** parameters. Back in Unity, starting with the **Add** button, follow these steps:

1. In the **Hierarchy** window, select the **Add** button under **UI Canvas | Main UI**.

2. From the **Hierarchy** window, drag the **Main Mode** game object (under **Interaction Controller**) onto the **Button | OnClick** action's **Object** slot.

3. In the **Function** selector, choose **Gallery Main Mode | Select Image To Add**.

4. Now, we'll wire up the **Replace Image** button, which is located under **UI Canvas | EditPicture UI | Edit Menu**.

5. In the **Inspector** window, on its **Button** component, click the + button at the bottom right of the **OnClick** actions.

6. From the **Hierarchy** window, drag the **EditPicture Mode** game object onto the **OnClick Object** slot.

7. In the **Function** selector, choose **Edit Picture Mode | Select Image To Replace**.

    The buttons are now set up. All we have to do now is assign the other references.

8. In the **Hierarchy** window, select the **Main Mode** game object (under **Interaction Controller**).

9. Drag the **SelectImage Mode** object from the **Hierarchy** window onto the **Select Image** slot.

10. In the **Hierarchy** window select the **SelectImage Mode** game object (under **Interaction Controller**).

11. Drag the **AddPicture Mode** object from the **Hierarchy** window onto the **Add Picture** slot.

12. Drag the **EditPicture Mode** object from the **Hierarchy** window onto the **Edit Picture** slot.

13. In the **Hierarchy** window, select the **EditPicture Mode** game object (under **Interaction Controller**).

14. Drag the **SelectImage Mode** object from the **Hierarchy** window onto the **Select Image** slot.

15. In the **Hierarchy** window, select the **Image Buttons** game object (under **UI Canvas | SelectImage UI**).

16. Drag the **SelectImage Mode** object from the **Hierarchy** window onto the **Select Image** slot.

That should do it!

In summary, we have refactored the `ImageButtons` script to call `SelectImageMode.ImageSelected` when a button is pressed. `SelectImageMode` will know whether the user is adding a new picture or replacing the image with an existing one. In the former case, the modal was called from Main-mode. In the latter case, the modal was called from EditPicture-mode and has an `isReplacing` flag set.

Go ahead and **Build and Run** the scene. Add a picture and then edit it. Then, tap the **Replace Image** button. The **Select Image** menu should appear. At this point, you can pick another image, and it will replace the one in the currently selected **FramedPhoto**. There are more features you could add to this project, including letting the user choose a different frame for their pictures.

Replacing the frame

The last **Edit** button we must implement is **Replace Frame**. I will leave this feature up to you to build since at this point, you may have the skills to work through this challenge on your own. A basic solution may be to keep the current **FramedPhoto** prefab and let the user just pick a different color for the frame. Alternatively, you could define separate frame objects within the **FramedPhoto** prefab, perhaps using models found on the Asset Store or elsewhere, and pick a frame that enables one or another frame object. Here are some suggestions regarding where to find models:

- *Classic Picture Frame*: `https://assetstore.unity.com/packages/3d/props/furniture/classic-picture-frame-59038`

- *Turbosquid*: `https://www.turbosquid.com/3d-model/free/picture-frame`

So far, we've been interacting with the placed object indirectly through the Edit menu buttons. Next, we'll consider directly interacting with the virtual object.

Interacting to edit a picture

We will now implement the ability to move and resize a virtual object we have placed in the AR scene. For this, I've decided to give the object being edited responsibility for its own interactions. That is, when **FramedPhoto** is being edited, it'll listen for input action events and move or resize itself.

I've also decided to implement these features as separate components, `MovePicture` and `ResizePicture`, on the **FramedPhoto** prefab. This will only be enabled while **FramedPhoto** is being edited. First, let's ensure that instantiated **FramedPhoto** objects receive Input Action messages so that they can respond to user input.

Ensuring FramedPhoto objects receive Input Action messages

We are currently using the Unity Input System, which lets you define and configure user input actions, as well as listening for those action events with a Player Input component. Currently, the scene has one Player Input component, attached to the Interaction Controller game object. The component is configured to broadcast messages down the local hierarchy. Therefore, if we want the `FramedPhoto` script to receive input action messages (which we now do), we must make sure the **FramedPhoto** object instances are children of the Interaction Controller. Let's simply parent the **FramedPhoto** objects under the **AddPicture Mode** game object where it's instantiated, as follows:

1. Edit the `AddPictureMode` script.

2. In the `PlaceObject` function, set the spawned object's parent as the **AddPicture Mode** game object by adding this line of code:

```
GameObject spawned = Instantiate(placedPrefab,
    position, rotation);
spawned.transform.SetParent(
    transform.parent);
```

The instantiated **FramedPhoto** prefabs will now be parented by the **AddPicture Mode** game object.

> **Information – Scene Organization and Input Action Messages**
>
> It's advisable to consider how you will organize your scene object hierarchy and where to place instantiated objects. For example, generally, I'd prefer to keep all our **FramedPhotos** in a separate root object container. If we did that now, we would have to set **Player Input Behavior** to invoke events, instead of broadcasting messages down the local hierarchy. And then, scripts responding to those input actions would subscribe (add listeners) to those messages (see `https://docs.unity3d.com/Packages/ com.unity.inputsystem@1.1/manual/Components. html#notification-behaviors`). On the other hand, for tutorial projects such as the ones in this book, I've decided that using the built-in input action messages is cleaner and more straightforward to explain.

Let's start by creating the empty scripts and adding them to the scene. Then, we'll build them out.

Adding the interaction components

To expedite the implementation, we must create the script files first by performing the following steps:

1. In your **Project** assets, create a new C# script named `MovePicture`.

2. Create another new C# script named `ResizePicture`.

3. Open the **FramedPhoto** prefab for editing.

4. Drag the **MovePicture** script and the **ResizePicture** script from the **Project** assets folder onto the root **FramedPhoto** object.

5. Edit the `FramedPhoto` script in your code editor. Add the following declarations at the top of the class:

```
MovePicture movePicture;
ResizePicture resizePicture;
```

6. Initialize it in `Awake` and start with the components disabled:

```
void Awake()
{
    movePicture = GetComponent<MovePicture>();
    resizePicture = GetComponent<ResizePicture>();
    movePicture.enabled = false;
    resizePicture.enabled = false;
```

```
        layer = LayerMask.NameToLayer("PlacedObjects");
        Highlight(false);
    }
```

7. Then, enable these components when editing:

```
    public void BeingEdited(bool editing)
    {
        Highlight(editing);
        movePicture.enabled = editing;
        resizePicture.enabled = editing;
        isEditing = editing;
    }
```

We've now prepared ourselves to add the move and resize direct manipulation features to the FramedPhoto object. These will be separate components that are enabled only while the picture is in EditPicture mode.

OK. Let's start by interactively moving the picture along the wall by dragging it with our finger on the screen.

Using our finger to move the picture

We will start by implementing the drag-to-move feature by adding a MoveObject action to the **AR Input Actions** asset. Like the **SelectObject** action (and **PlaceObject**) that we already have, this will be bound to the touchscreen's primary touch position. We'll keep this action separate from the others, for example, should you decide to use a different interaction technique, such as a touch and hold, to start the dragging operation. But for now, we can just copy the other one, as follows:

1. In the **Project** window, *double-click* the **AR Input Actions** asset (in the Assets/ Inputs/ folder) to open it for editing (or use its **Edit Asset** button).

2. In the middle section, *right-click* the **SelectObject** action and select **Duplicate**.

3. Rename the new one MoveObject.

4. Press **Save Asset** (unless **Auto-Save** is enabled).

Now, we can add the code that will listen for this action. Edit the MovePicture script and write the following:

```
using System.Collections.Generic;
using UnityEngine;
```

```csharp
using UnityEngine.EventSystems;
using UnityEngine.InputSystem;
using UnityEngine.XR.ARFoundation;
using UnityEngine.XR.ARSubsystems;

public class MovePicture : MonoBehaviour
{
    ARRaycastManager raycaster;
    List<ARRaycastHit> hits = new List<ARRaycastHit>();

    void Start()
    {
        raycaster = FindObjectOfType<ARRaycastManager>();
    }

    void Start(){ }

    public void OnMoveObject(InputValue value)
    {
        if (!enabled) return;
        if (EventSystem.current.IsPointerOverGameObject(0))
            return;

        Vector2 touchPosition = value.Get<Vector2>();
        MoveObject(touchPosition);
    }

    void MoveObject(Vector2 touchPosition)
    {
        if (raycaster.Raycast(touchPosition, hits,
            TrackableType.PlaneWithinPolygon))
        {
            ARRaycastHit hit = hits[0];
            Vector3 position = hit.pose.position;
            Vector3 normal = -hit.pose.up;
```

```
        Quaternion rotation =
            Quaternion.LookRotation(normal, Vector3.up);
        transform.position = position;
        transform.rotation = rotation;
    }
  }
}
```

This code is very similar to that in the AddPictureMode script. It's using **AR Raycast Manager** to find a trackable plane and place the object so that it's flush with the plane and upright. The difference is that we're not instantiating a new object, we're just updating the transform of the existing one. And we're doing this continuously, so long as the input action events are being generated (that is, so long as the user is touching the screen).

The OnMoveObject function is skipped if the input action message is received but this component is not enabled. It also checks that the user is not tapping a UI element (an event system object), such as one of our edit menu buttons.

Try it out. If you play the scene, create a picture, and begin editing it, you should be able to drag the picture with your finger and it will move along the wall plane. In fact, since we are raycasting each update, it could find a newer, refined tracked plane as you're dragging, or even move the picture to a different wall.

As we mentioned previously, if you tap the screen on any tracked plane, the current picture will "jump" to that location. If that is not your desired behavior, we can check that the initial touch is on the current picture before we start updating the transform position. The modified code is as follows:

1. Declare and initialize references to camera and layerMask:

```
Camera camera;
int layerMask;

void Start() {
    raycaster = FindObjectOfType<ARRaycastManager>();
    camera = Camera.main;
    layerMask =
        1 << LayerMask.NameToLayer("PlacedObjects");
}
```

2. Add a raycast to `MoveObject` to ensure the touch is on a picture before you move it:

```
void MoveObject(Vector2 touchPosition)
{
    Ray ray = camera.ScreenPointToRay(touchPosition);
    if (Physics.Raycast(ray, Mathf.Infinity, layerMask))
    {
        if (raycaster.Raycast(touchPosition, hits,
            TrackableType.PlaneWithinPolygon))
        {
            ARRaycastHit hit = hits[0];
            Vector3 position = hit.pose.position;
            Vector3 normal = -hit.pose.up;
            Quaternion rotation =
                Quaternion.LookRotation(normal,
                    Vector3.up);
            transform.position = position;
            transform.rotation = rotation;
        }
    }
}
```

Currently, we only have the tracked planes visible in AddPicture-mode. I think it would be useful to also show them in Edit-mode. We can use the same `ShowTrackablesOnEnable` script we wrote in a previous chapter that's already been applied to the **AddPicture Mode** game object. Add this as follows:

1. In the **Hierarchy** window, select the **EditPicture Mode** game object (under **Interaction Controller**).

2. Locate the `ShowTrackablesOnEnable` script in your Project `Scripts/` folder.

3. From the **Hierarchy** window, drag the **AR Session Origin** game object onto the **Show Trackables On Enable | Session Origin** slot.

4. Drag the script onto the **EditPicture Mode** object, adding it as a component.

Now, when **EditPicture Mode** is enabled, the trackable planes will be displayed. When it's disabled and you go back to Main-mode, they'll be hidden again.

Next, we'll implement the pinch-to-resize feature.

Pinching to resize the picture

To implement pinch-to-resize, we'll also use an Input Action, but this will require a two-finger touch. As such, the action is not simply returning a single value (for example, Vector2). So, this time, we'll use a **PassThrough** Action Type. Add it by performing the following steps:

1. Edit the **AR Input Actions** asset, as we did previously.

2. In the middle **Actions** section, select + and name it `ResizeObject`.

3. In the rightmost **Properties** section, select **Action Type | Pass Through**, and **Control Type | Vector 2**.

4. In the middle **Actions** section, select the **<No Binding>** child. Then, in the **Properties** section, select **Properties | Path | Touchscreen | Touch #1 | Position** to bind this action to a second finger screen touch.

5. Press **Save Asset** (unless **Auto-Save** is enabled).

Now, we can add the code to listen for this action. Edit the `ResizePicture` script and write it as follows. In the first part of the script, we declare several properties that we can use to tweak the behavior of the script from the Unity Inspector. `pinchspeed` lets you adjust the sensitivity of the pinch, while `minimumScale` and `maximumScale` let you limit how small or big the user will end up making the picture, respectively. Follow these steps:

1. Begin the script with the following code:

```
using UnityEngine;
using UnityEngine.EventSystems;
using UnityEngine.InputSystem;


public class ResizePicture : MonoBehaviour
{
    [SerializeField] float pinchSpeed = 1f;
    [SerializeField] float minimumScale = 0.1f;
    [SerializeField] float maximumScale = 1.0f;


    float previousDistance = 0f;


    void Start() { }
```

Note that I declared an empty `Start()` function. This is needed because a `MonoBehaviour` component without a `Start` or `Update` function cannot be disabled (you'll see this for yourself if you remove `Start` from the code and look at it in the **Inspector** window – you'll see that the **Enable** checkbox is missing).

2. The `OnResizeObject` function is the listener for the input action messages. Because we specified the Action Type as **Pass Through**, there are no incoming arguments to the function. Instead, we can read the current state of `Touchscreen` to get the first and second finger touches. Then, we can pass those touch positions to our `TouchToResize` function:

```
public void OnResizeObject()
{
    if (!enabled) return;
    if (EventSystem.current.
        IsPointerOverGameObject(0)) return;

    Touchscreen ts = Touchscreen.current;
    if (ts.touches[0].isInProgress &&
        ts.touches[1].isInProgress)
    {
        Vector2 pos =
            ts.touches[0].position.ReadValue();
        Vector2 pos1 =
            ts.touches[1].position.ReadValue();
        TouchToResize(pos, pos1);
    }
    else
    {
        previousDistance = 0;
    }
}
```

3. The `TouchToResize` algorithm is straightforward. It gets the distance between the two finger touches (in screen pixels) and compares it against the previous distance. Dividing the new distance by the previous distance gives us the percentage change, which we can use to directly modify the transform scale. It seems to work pretty well for me:

```
void TouchToResize(Vector2 pos, Vector2 pos1)
{
    float distance = Vector2.Distance(pos, pos1);

    if (previousDistance != 0)
    {
        float scale = transform.localScale.x;
        float scaleFactor = (distance /
            previousDistance) * pinchSpeed;
        scale *= scaleFactor;
        if (scale < minimumScale)
            scale = minimumScale;
        if (scale > maximumScale)
            scale = maximumScale;
        Vector3 localScale = transform.localScale;
        localScale.x = scale;
        localScale.y = scale;
        transform.localScale = localScale;
    }
    previousDistance = distance;
}
```

Try it out. If you play the scene, create a picture, and begin editing it, you should be able to use two fingers to resize the picture, pinching your fingers together to make it smaller and un-pinching them apart to increase the picture's size. Here's a screen capture from my phone with some pictures arranged on my dining room wall, all of which are various sizes:

Figure 7.7 – Virtual framed photos arranged on my dining room wall

In this section, we looked at how to directly interact with virtual objects. Using input actions, we added features using the touchscreen to drag and move a picture on a wall, as well as pinching to resize a picture.

We could improve this by adding a Cancel Edit feature that restores the picture to its pre-edited state. One way to do this is to make a temporary copy of the object when it enters edit mode, and then restore or discard it if the user cancels or saves their changes, respectively.

Another feature worth considering is persisting the picture object arrangements between sessions, so that the app saves your pictures when you exit the app and restores them when you restart the app. This is an advanced topic that I will not cover in this book since it is outside of Unity AR Foundation itself. Each provider has its own proprietary solutions. If you're interested, take a look at *ARCore Cloud Anchors*, which is supported by Unity *ARCore Extensions* (`https://developers.google.com/ar/develop/unity-arf/cloud-anchors/overview`) and *ARKit ARWorldMap* (`https://developer.apple.com/documentation/arkit/arworldmap`), as exposed in the Unity *ARKit XR Plugin* (`https://docs.unity3d.com/Packages/com.unity.xr.arkit@4.0/api/UnityEngine.XR.ARKit.ARWorldMap.html`).

This concludes our exploration of, and building, an AR photo gallery project.

Summary

In this chapter, you expanded on the AR gallery project we began in *Chapter 6, Gallery: Building an AR App*. That project left us with the ability to place framed photos on our walls. In this chapter, you added the ability to edit virtual objects in the scene.

You implemented the ability to select an existing virtual object in Main-mode, where the selected object is highlighted and the app goes into EditPicture-mode. Here, there is an edit menu with buttons for **Replace Image**, **Replace Frame**, **Remove Picture**, and **Done** (return to Main-mode). The **Replace Image** feature displayed the same **SelectImage** modal menu that is used when we're creating (adding) new pictures. We had to refactor the code to make it reusable.

While placing and moving a picture on the wall, you implemented a feature to avoid overlapping or colliding objects, automatically moving the picture away from the other ones. After that, you implemented some direct interactions with the virtual objects by using touch events to drag a picture to a new location. You also implemented pinching to resize pictures on the wall. Finally, you learned how to use more Unity APIs from C#, including collision trigger hooks and vector geometry.

In the next chapter, we'll begin a new project while using a different AR tracking mechanism – tracked images – as we build a project for visualizing 3D data; namely, the planets in our Solar System.

8
Planets: Tracking Images

In this chapter, we will be using augmented reality for data visualization and education. We're going to build a project where users can learn about the planets in our Solar System. Suppose you have a children's science book on the Solar System with a companion mobile app. On the page about planet Earth, for example, the reader can point their mobile device at the picture on the page and a 3D rendering of the Earth will pop out of the page.

The AR mechanism we'll be using is known as *image tracking*. With image tracking, you prepare a reference library of images that may be recognized and tracked in the real world at runtime. When the user's device's camera detects one of these images, a virtual object can be instantiated at the image location.

I have provided you with "planet cards," which have pictures and unique markers on them for each planet that I created from free resources available on the web, for you to print yourself and use with the app. For rendering the planets' spherical surface skins, we will be using free texture images of the actual planets.

We will cover the following topics in this chapter:

- Understanding AR image tracking
- Specifying the Planets project and getting started
- Defining and tracking reference images

- Creating and instantiating a virtual Earth prefab
- Rotating a planet on its axis
- Expanding the project with multiple planets
- Making a responsive UI

By the end of this chapter, you'll have a working app that detects images on the provided planet cards, renders a 3D model of the given planet, and offers additional information details about a planet.

Technical requirements

To implement the project in this chapter, you need Unity installed on your development computer, connected to a mobile device that supports augmented reality applications (see *Chapter 1, Setting Up for AR Development*, for instructions). We also assume that you have the ARFramework template and all its prerequisites installed. See *Chapter 5, Using the AR User Framework*, for more details. The completed project can be found in this book's GitHub repository at https://github.com/PacktPublishing/Augmented-Reality-with-Unity-AR-Foundation.

Understanding AR image tracking

Before we start building our project, let's take a moment to learn how AR image tracking works. In this section, I'll introduce some of the basic principles behind image recognition and tracking, and what makes some images better than others for this purpose.

As we know, the principles behind augmented reality involve using compute mechanisms to recognize features in the real world, determine their position and orientation in a 3D space, instantiate virtual objects relative to and anchored within this 3D space, and track the user as they move within this space. Modern devices can accomplish this using their video cameras and other sensors built into the device to performing real-time spatial mapping of the environment. A different approach is for the device to track predetermined images. That is what we will use for the project in this chapter.

Augmented reality technology was born in the 1990s, where QR code-like marker images were used for tracking. An example is shown in the following image:

Figure 8.1 – A basic AR marker

Marker images can be used for triggering and positioning virtual objects in the real world. These simplistic yet visually distinct markers are easily detected, even by low-end devices. Such markers are readily detectable because of their *distinctive details*, *high contrast edges*, and an *asymmetric shape* – that is, it's an easily recognizable image with unambiguous top, bottom, left, and right sides. In this way, the detection software can determine which marker image is in view and the orientation of the camera relative to the marker in 3D space.

Taken to the next level, products such as Merge Cube (`https://mergeedu.com/cube`) have markers on each of its six faces, just like a physical cube that you can hold in your hand. Users can find companion apps with a wide gamut of games, learning, and exploration experiences. Merge offers a Unity package for developers so that you can build your own projects for it too. Merge Cube is depicted in the following image:

Figure 8.2 – Merge Cube provides a 3D tracking cube with markers on each face

Markers can be combined with natural images to provide pleasing and informative yet visually distinct images that also act as AR markers. You'll often see this in AR augmented storybooks or even cereal boxes. This is the approach I have taken in this chapter.

While markers provide the highest reliability, they are not necessarily required for image tracking. Ordinary photographic images can also be used. In AR lingo, these are referred to as *natural feature images*. Images for tracking must have the same characteristics that make markers reliable – distinctive details, high contrast edges, and an asymmetric shape. Much has been written about the best practices for selecting images. For instance, the AR Core developer guide (`https://developers.google.com/ar/develop/java/augmented-images/`) contains additional tips about using reference images, including the following:

- Use an image resolution of at least 300 x 300 pixels. However, a very high resolution does not help with recognition.

- Color information is not used, so either color or grayscale images are just as good.

- Avoid images with a lot of geometric features, or too few.

- Avoid repeating patterns.

The AR Core SDK comes with an **arcoreimg** tool that can evaluate images and returns a quality score between 0 and 100 for each image, where a score of at least 75 is recommended. Likewise, Unity uses a similar tool when compiling the Image Reference Library in your builds (we'll learn more about this later in this chapter).

Given this general understanding of using image tracking in augmented reality applications, let's begin by defining a fun and interesting project – visualizing our Solar System's planets.

Specifying the Planets project

We are going to build a planet information app that allows users to scan *planet cards* to visualize a 3D model of each planet in the Solar System. Imagine this being part of a trading card collection or a companion app to a children's science book. When the user points the device's camera at one of the planet cards, they can see a 3D rendering of the planet. Upon pressing an **Info** button, the user can get additional information about that planet. In this section, I will define the general user experience flow, give you instructions for preparing the planet cards for your own use, and help you collect assets that you'll use in this project.

User experience flow

The general onboarding user workflow will play out as follows:

1. **Startup-mode**: The app will start, check the device for AR support, and ask for camera permissions (OS-dependent). Once read, the app will go into Scan-mode.

2. **Scan-mode**: The user is prompted to aim the camera at an image for detection and tracking. When at least one image is being tracked, the app goes into Main-mode.

3. **Main-mode**: This is where the app responds to new or updated tracked images and allows the user to interact with the planet. When an image is tracked, it determines which planet corresponds to the image and instantiates the planet's game object. If tracking is lost, the app may go back to Scan-mode to prompt the user. If a different image is tracked, the current planet is replaced with the new image's planet.

This workflow is a bit simpler than the ones we implemented in the previous chapters. In that case, we needed the user to scan the environment for trackable planes before starting Main-mode. The user was then asked to deliberately tap the screen to place a virtual object in the scene. Furthermore, in the AR Gallery project, we added Edit-mode to modify pictures that had been added by the user. Much of that is unnecessary in this project; the process is more automated as we let the device detect an image and we instantiate a virtual object in response.

Preparing the planet cards

For this project, we are using printed *planet cards* as marker images so that we can choose a planet to visualize. You can find a PDF file that contains the cards in the project files for this chapter (in the folder named `Printables/`). To prepare the cards for this project, follow these steps:

1. Print out the `PlanetCards.pdf` file.

2. Then, cut the sheets into separate cards.

3. I suggest that you print on thick paper stock or mount the printouts on paperboard to avoid warping, which may affect the software's ability to recognize the images at runtime.

The following photo shows getting these cards ready for use:

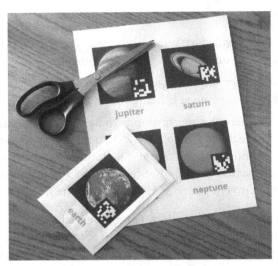

Figure 8.3 – Cutting the printed planet cards for this project

These cards were created from a combination of resources that can be found for free on the web. I found the original flashcards on the *Kids Flashcards* website. Upon going to `https://kids-flashcards.com/en/free-printable/solar-system-flashcards-in-english`, I downloaded the *Solar System flashcards free* PDF file.

First, I attempted to use the flashcards as-is, but the pictures were not distinctive enough to be detected individually. So, I decided to add *ArUco* markers to each one. ArUco is a square marker with a wide black border and inner binary matrix that determines its ID based on OpenCV (the Open Source Computer Vision library, which was developed at the *University of Cordoba, Spain*; see `https://docs.opencv.org/3.2.0/d5/dae/tutorial_aruco_detection.html`). I used the online ArUco marker generator at `https://chev.me/arucogen/` to make separate markers for each planet.

Then, I used *Photoshop* to combine the markers with the planet flashcards to make our final planet cards for this project. (The Photoshop PSD file is also included with this chapter's files on GitHub.)

Each planet card is also a separate PNG image. These have been provided for you in the `Image Library/` folder. Later in this chapter, we will create an image reference library and add these images. The images are named with the pattern *[planetname]*-`MarkerCard.png`; for example, `Earth-MarkerCard.png`. We'll take advantage of this naming convention in our code.

When the app detects a planet card, the application will instantiate a model of the planet. For this, we need texture images for the planet materials.

Collecting planet textures and data

We need texture images to use as the planet skins of the spherical mesh for each planet. The ones we're using I found at the interesting *Solar System Scope* project site (`https://www.solarsystemscope.com/`). These are included with the files for this chapter in this book's GitHub repository and can be downloaded from `https://www.solarsystemscope.com/textures/`. That said, you can find alternative assets in the Unity Asset Store (`https://assetstore.unity.com/?q=solar%20system&orderBy=1`), including the classic *Planet Earth Free* package (`https://assetstore.unity.com/packages/3d/environments/sci-fi/planet-earth-free-23399`) for Earth itself, which includes cloud cover.

For additional metadata about the planets, I found the *Planetary Fact Sheet* on the NASA.gov website (`https://nssdc.gsfc.nasa.gov/planetary/factsheet/index.html`) and more details at `https://nssdc.gsfc.nasa.gov/planetary/planetfact.html`. We could use some of these details directly while rendering and animating our models, such as the planet diameter (km), rotation period (hours), and tilt (obliquity to orbit in degrees).

With our planet cards, planet skin textures, and other planetary details in hand, we're ready to start building the project.

Getting started

To begin, we'll create a new scene named `PlanetsScene` using the `ARFramework` scene template. Follow these steps:

1. Select **File | New Scene**.

2. In the **New Scene** dialog box, select the **ARFramework** template.

3. Press **Create**.

4. Select **File | Save As**. Navigate to the `Scenes/` folder in your `Assets` project, name it `PlanetsScene`, and click **Save**.

The new AR scene already has the following set up:

- **AR Session** game object.

- **AR Session Origin** rig with the raycast manager and plane manager components.

- **UI Canvas** is a screen space canvas with child panels; that is, Startup UI, Scan UI, Main UI, and NonAR UI. It also contains the UI Controller component script that we wrote.

- **Interaction Controller** is a game object that contains the Interaction Controller component script we wrote, which helps the app switch between interaction modes, including the Startup, Scan, Main, and NonAR modes. It also has a **Player Input** component that's been configured with the **AR Input Actions** asset we created previously.

- The **OnboardingUX** prefab from the AR Foundation Demos project, which provides AR session status and feature detection status messages, as well as animated onboarding graphics prompts.

We can set the app title now, as follows:

1. In the **Hierarchy** window, unfold the **UI Canvas** object and unfold its child **App Title Panel**.

2. Select the **Title Text** object.

3. In its **Inspector** window, change its text content to `Planet Explorer`.

Using this scene as a basis, we will replace the AR trackable components with an AR Tracked Image Manager one.

Tracking reference images

Our starter scene includes an AR Session Origin with components for Player Input and AR Raycast Manager. It also has a component we do not need in this project, for detecting and tracking planes, which we'll replace with AR Tracked Image Manager instead. Documentation on **AR Tracked Image Manager** can be found at `https://docs.unity3d.com/Packages/com.unity.xr.arfoundation@4.1/manual/tracked-image-manager.html`. Then, we'll create an image reference library for our planet card images.

Adding AR Tracked Image Manager

To configure the AR Session to track images, perform the following steps:

1. In the **Hierarchy** window, select the **AR Session Origin** game object.

2. In the **Inspector** window, use the **3-dot** context menu (or *right-click*) on **AR Plane Manager** and select **Remove Component**.

3. Using the **Add Component** button, search for AR and add an **AR Tracked Image Manager** component.

You'll notice that there is a **Serialized Library** slot on the component for the reference image library. We'll create that next.

Creating a reference image library

The reference image library contains records for each of the images that the application will be able to detect and track in the real world. In our case, we're going to add the planet card images. In the assets provided in the GitHub repository for this book, there is a folder named Image Library/ that already contains the planet card images we'll be adding to the library. We will start with just the Earth card here; we will add the other planets later in this chapter.

We can create the library by performing the following steps:

1. In the **Project** window, find or create a folder named Image Library/.
2. *Right-click* on the Image Library/ folder and select **Create | XR | Reference Image Library**.
3. With the **ReferenceImageLibrary** assets selected, in the **Inspector** window, click **Add Image**.
4. Drag the Earth-MarkerCard image from the **Project** window onto the square image texture slot.
5. Check the **Specify Size** checkbox.
6. If you printed the planet cards from the PDF provided, at scale, the width will be about 8 cm, or 0.08 meters. Otherwise, use a ruler to measure the Earth planet card you printed.
7. Then, enter the width (0.08) in the **X** field. The **Y** value will be automatically updated based on the PNG image's pixel dimensions.
8. Check the **Keep Texture At Runtime** checkbox.

The resulting **Reference Image Library** settings are shown in the following screenshot:

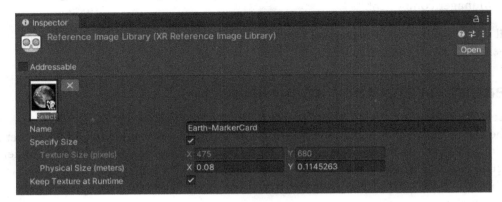

Figure 8.4 – Reference Image Library with the Earth added

Now, we can update the **AR Tracked Image Manager** component, as follows:

1. In the **Hierarchy** window, select the **AR Session Origin** object.

2. Drag the **ReferenceImageLibrary** asset from the **Project** window onto the **Serialized Library** slot of **AR Tracked Image Manager**.

3. Temporarily, while we get this project set up, we'll instantiate an existing prefab object when the image is detected.

    For example, drag the **AR Placed Cube** prefab from the ARF-samples/ Prefabs/ folder onto the **Tracked Image Prefab** slot (or another similar object).

The **AR Tracked Image Manager** component should now look as follows:

Figure 8.5 – The AR Tracked Image Manager with the reference image library assigned

You now have an AR scene that recognizes and tracks images that have been defined in a reference library. Currently, the library only contains the Earth-MarkerCard image. When the image is recognized while running the app, a simple cube will be placed on the Earth planet card.

We're almost ready to try this out. But first, let's configure the user framework's UI and modes.

Configuring the user interaction modes and UI

The scene template, ARFramework, where we started provides a simple framework for enabling user interaction modes and displaying the corresponding UI panels for a mode. This project will start in Startup-mode while the AR Session is initializing so that we can verify that the device supports AR. Then, it will transition to Scan-mode, where it will try to find one of the reference images. Once found, it will transition to Main-mode, where we can support additional user interactions with the app's content.

Scanning for reference images

In Scan-mode, we'll display an instructional graphic prompting the user to point the camera at a planet card with a planet and marker image. Perform the following steps to configure this:

1. In the **Hierarchy** window, unfold the **UI Canvas** game object and unfold its child **Scan UI**. Select the child **Animated Prompt** object.

2. In the **Inspector** window set **Animated Prompt | Instruction to Find An Image**.

This will now play the Find Image Clip we defined on the **OnboardingUX** object, which is provided by the Unity Onboarding UX assets and is already present in our scene hierarchy. What you can expect is shown in the following screen capture. On the left is Startup-mode, where the AR Session is being initialized. On the right is Scan-mode, where the user is prompted to find an image (you can't see the video feed because I'm covering the camera to make the prompt more visible in the screen capture).

Figure 8.6 – Screen captures of Startup mode (left) and Scan mode (right)

Now, we need to set up the Scan-mode's script to know when an image has been found and transition to Main-mode. We'll replace the default `ScanMode` script with a similar one that references `ARTrackedImageManager` instead of `ARTrackedPlaneManager`, as follows:

1. In the **Project** window, create a new C# script in your `Scripts/` folder by *right-clicking* and selecting **Create | C# Script**. Name the new file `ImageScanMode`.

2. Edit `ImageScanMode` and replace its content, as follows:

```
using UnityEngine;
using UnityEngine.XR.ARFoundation;


public class ImageScanMode : MonoBehaviour
{
    [SerializeField] ARTrackedImageManager imageManager;

    private void OnEnable()
    {
        UIController.ShowUI("Scan");
    }

    void Update()
    {
        if (imageManager.trackables.count > 0)
        {
            InteractionController.EnableMode("Main");
        }
    }
}
```

3. Save the script. Then, back in Unity, in the **Hierarchy** window, select the **Scan Mode** game object (under **Interaction Controller**).

4. In the **Inspector** window, remove the original **Scan Mode** component using the *3-dot* context menu and selecting **Remove Component**.

5. Drag the **ImageScanMode** script onto the **Scan Mode** object, adding it as a new component.

6. From the **Hierarchy** window, drag the **AR Session Origin** object into the **Inspector** window and drop it onto the **Image Scan Mode | Image Manager** slot.

The component will now look as follows:

Figure 8.7 – The Image Scan Mode component

Currently, we have created a new scene using the ARFramework template and modified it to use **AR Tracked Image Manager** and prompt the user to scan for an image accordingly. When an image is detected (for example, the Earth-MarkerCard), a generic game object will be instantiated (for example, the AR Placed Cube prefab). Let's test what we have accomplished so far on the target device.

Build and run

To build and run the scene on your target device, perform the following steps:

1. Ensure you've saved the work you've done on the current scene by going to **File | Save**.

2. Select **File | Build Settings** to open the **Build Settings** window.

3. Click **Add Open Scenes** to add the current scene to the **Scenes In Build** list (if it's not already present).

4. Uncheck all but the current scene, PlanetsScene, from the list.

5. Then, click **Build And Run** to build the project.

When the app launches, point your device's camera at the printed Earth planet card. Your virtual cube should get instantiated at that location, as shown in the following screen capture from my phone:

Figure 8.8 – The Earth card has been detected, and the cube has been instantiated

We now have a basic AR scene with image detection set up to recognize the Earth planet card and instantiate a sample prefab at that location. Now, let's make a planet Earth model that we can use instead of this silly cube.

Creating and instantiating a virtual Earth prefab

In this section, we will create prefab game objects for each of the planets. Since each of the planets has similar behaviors (for example, they rotate), we'll first create a generic Planet Prefab, and then make each specific planet a variant of that one. In Unity, **prefab variants** allow you to define a set of predefined variations of prefabs, such as our planet one (see `https://docs.unity3d.com/Manual/PrefabVariants.html`). We'll write a `Planet` script that animates the planet's rotation and handles other behavior. Each planet will have its own "skin" defined by a material, along with a base texture map, which we downloaded earlier from the web.

In this section, we'll create a generic Planet Prefab object, create an Earth Prefab as a variant, add planet metadata by writing a `Planet` component script, and implement a planet rotation animation.

Creating the generic Planet Prefab

The Planet Prefab contains a 3D sphere that gets rendered with each planet's texture image. Planets spin along their axes, so we'll set up a hierarchy with an Incline transform that defines this incline axis. Follow these steps:

1. In your **Project** window, *right-click* and select **Create | Prefab** (create the folder first if necessary). Name it `Planet Prefab`.

2. *Double-click* (or select **Open Prefab** in the **Inspector** window) to open the prefab for editing.

3. From the main menu, select **GameObject | Create Empty** and name it `Incline`.

4. Right-click the **Incline** game object in the **Hierarchy** window and select **3D Object | Sphere**. Name it `Planet`.

5. It will be useful to have any planets we instantiate in the scene on a specific layer. I will name this layer `PlacedObjects`. (I introduced and discussed layers in a previous chapter). With its root **Planet Prefab** object selected, in the top right of its **Inspector** window, click the **Layer** drop-down list and select `PlacedObjects`.

    If the `PlacedObjects` layer doesn't exist, select **Add Layer** to open the **Layers manager** window. Add the name `PlacedObjects` to one of the empty slots. In the **Hierarchy** window, click the **FramedPhoto Prefab** object to get back to its **Inspector** window. Again, using the **Layers** drop-down list, select `PlacedObjects`.

You will then be prompted with the question, **Do you want to set layer to PlacedObjects for all child objects as well?** Click **Yes, Change Children**.

6. **Save** the prefab.

This is very simplistic right now (only a sphere child object is being parented by an Incline transform), but it will serve as a template for each planet prefab that we add. The Planet Prefab hierarchy is shown in the following screenshot:

Figure 8.9 – The Planet Prefab hierarchy

Each planet will be rendered with a skin representing an actual view of that planet. Before creating the Earth prefab, let's take a moment to understand render materials and the texture images we are going to use.

Understanding equirectangular images

When Unity renders a 3D model, it starts with a 3D mesh that describes the geometry. Much like a fishing net, a mesh is a collection of vertices and vectors, with the vectors connecting these vertices, organized as triangles (or sometimes four-sided quads) that define the surface of the mesh. The following illustration shows a wireframe view of a sphere mesh on the left. On the right is a rendered view of the sphere, with a globe texture mapped onto its 3D surface:

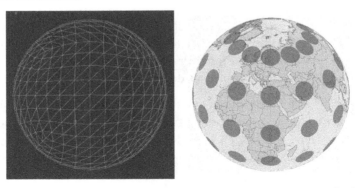

Figure 8.10 – Sphere mesh (left) and rendered sphere with texture (right)

A texture image is just a 2D image file (for example, a PNG file) that is computationally mapped onto the 3D mesh's surface when it is rendered. Think of unraveling a globe as a 2D map, like cartographers have been doing for centuries. A common 2D projection is known as **equirectangular**, where the center (equator) is at the correct scale and the image gets increasingly stretched as it approaches the top and bottom poles. The following image shows the equirectangular texture of the preceding globe (illustration by *Stefan Kuhn*):

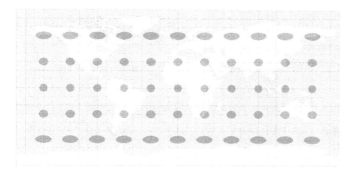

Figure 8.11 – Equirectangular texture that defines the skin of a sphere

Information – Equirectangular Images Are Also Used in 360-Degree Media and VR

Equirectangular images are also known as 360-degree images and used in virtual reality applications. In VR, the image is effectively mapped to the *inside* of a sphere, where you're viewing from the inside rather than the outside of a globe!

For our project, we have texture images for each of the planets. The Mars one, for example, is as follows:

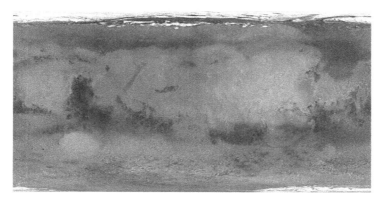

Figure 8.12 – Texture map image for Mars

To create a prefab for a specific planet, such as Earth, we'll need to create a material that uses the Earth texture image. We'll build that now.

Creating the Earth prefab

The Earth prefab will be a variant of the Planet Prefab, with its own Earth Material. Create it by performing the following steps:

1. In the **Project** window, *right-click* **Planet Prefab** and select **Create | Prefab Variant**. Name the new asset `Earth Prefab`.

2. *Double-click* **Earth Prefab** (or select **Open Prefab** in the **Inspector** window).

3. In the **Project** window, *right-click* in your `Materials/` folder (create one if necessary) and select **Create | Material**. Rename it `Earth Material`.

4. Drag **Earth Material** from the **Project** window and drop it onto the **Planet** game object.

5. Locate the **Earth** texture image (for example, in `Planet Textures/earth`) in the **Project** Assets and drag it onto the **Surface Inputs | Base Map** texture chip. The following screenshot shows **Earth Material** in the **Inspector** window:

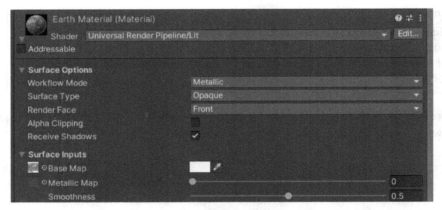

Figure 8.13 – Earth Material with the Base Map texture defined

6. Let's pick a default size for our planet when added to the scene. Unless you want to place a 1-meter diameter planet into your scene (!), we need to set its **Scale**.

    Select the **Planet** child object in the **Hierarchy** window.

7. In its **Inspector** window, set **Transform | Scale | X, Y Z** to (`0.1, 0.1, 0.1`).

8. Likewise, to rest the planet on the image's surface, we could set its Y position to 0.05. But to let it hover a little above, we will set **Transform | Position | Y** to 0.075.

9. **Save** the prefab and exit back to the **Scene** hierarchy.

Use this prefab instead of the AR Placed Cube prefab in the AR Tracked Image Manager component on the AR Session Origin object. Later, we'll manage this more correctly using a script but for now, let's just try it out:

1. In the **Hierarchy** window, select the **AR Session Origin** game object.

2. Drag **Earth Prefab** from the **Project** window into the **Inspector** window and drop it into the **AR Tracked Image Manager | Tracked Image Prefab** slot.

3. **Build and Run** the scene.

This time, when you point the camera at the Earth planet card, the Earth prefab will appear, as shown in the following screen capture:

Figure 8.14 – While tracking the Earth planet card, the app instantiates an Earth prefab

This looks pretty nice. The prefab could also include other information about the planet. We'll look at how to do this next.

Adding planet metadata

Each planet prefab can include additional information about that planet. We can capture this in the `Planet` script of the prefab, as follows:

1. From the **Project** window, open **Planet Prefab** for editing.

2. In the **Project** window's `Scripts/` folder, create a new C# script named `Planet`.

3. Drag the `Planet` script onto the root **Planet Prefab** game object, adding it as a component.

4. Open the `Planet` script in your code editor and write the following:

```
using UnityEngine;
public class Planet : MonoBehaviour
{
    public string planetName;
    public string description;
}
```

5. Save the script. Then, in Unity, **Save** the prefab.

   Although we have made all these changes to the Planet Prefab, the Earth Prefab inherits everything because it is a prefab variant.

6. Now, open **Earth Prefab** for editing.

7. In the **Planet Name** field, enter `Earth`.

8. In the **Description** field, enter a text description that we'll use later in the project, such as `Earth is the third planet from the Sun and the only astronomical object known to harbor and support life`.

9. **Save** the prefab.

We also can ascribe behaviors to the planet prefab, such as rotation about its axis.

Animating the planet's rotation

Planets spin. Some faster than others. Mercury just barely – it rotates once every 59 Earth days, while it orbits the Sun in 88 Earth days! And a planet's axis of rotation is not perfectly vertical (relative to its orbit around the Sun). Earth, for example, is tilted by 23.4 degrees, while Venus rotates on its side at 177.4 degrees! OK, enough science trivia – let's animate our Earth model. We're going to add a `Planet` behavior script to the Planet Prefab that rotates the planet along its rotation axis. Follow these steps to do so:

1. Open the `Planet` script in your code editor and add the following code:

    ```
    [SerializeField] private float inclineDegrees =
        23.4f;
    [SerializeField] private float rotationPeriodHours =
        24f;
    [SerializeField] private Transform incline;
    [SerializeField] private Transform planet;
    public float animationHoursPerSecond = 1.0f;

    void Start()
    {
        incline.Rotate(0f, 0f, inclineDegrees);
    }

    void Update()
    {
        float speed =
            rotationPeriodHours * animationHoursPerSecond;
        planet.Rotate(0f, speed * Time.deltaTime, 0f);
    }
    ```

At the top of the class, we will declare variables for `inclineDegrees` (Earth is 23.4) and `rotationPeriodHours` (Earth is 24). We will also define references to the prefab's `incline` and `planet` child objects.

There's also a public `animationHoursPerSecond`, which sets the animation speed. I've initialized it to `1.0`, which means the Earth will complete one rotation in 24 seconds.

The `Start()` function sets up the **Incline** angle by rotating along the Z-axis. This only needs to be done once.

The `Update()` function rotates the planet about its local Y-axis. Since the planet is parented by the **Incline** transform, it appears to rotate about the tilted incline axis. Multiplying the speed by `Time.deltaTime` each `Update` is a common Unity idiom for calculating how an object's Transform changes from one frame to the next, where `deltaTime` is the fraction of a second since the previous `Update`.

After saving the script, back in Unity, do the following:

1. From the **Project** window, open **Planet Prefab** for editing.
2. Ensure the root **Plane Prefab** game object is selected in the **Hierarchy** window.
3. Drag the **Incline** game object from the **Hierarchy** window into the **Inspector** window before dropping it onto the **Planet | Incline** slot.
4. Drag the **Planet** object onto the **Planet | Planet** slot.

The **Planet** component will now look like this in the **Inspector** window:

Figure 8.15 – The Planet component on the Planet Prefab

Please now **Build and Run** the project. When the Earth is instantiated, it will be tilted and rotating at the rate of one full rotation every 24 seconds.

At this point, we have a basic AR scene with image tracking. It lets the AR Tracked Image Manager instantiate our Earth Prefab directly when an image is detected. Currently, it doesn't distinguish what image is detected (supposing you had multiple images in the reference library) and always instantiates an Earth Prefab. We need to make the app more robust, and we can do this from the Main-mode.

Building the app's Main-mode

As you now know, **AR Tracked Image Manager** (on the **AR Session Origin** game object) performs 2D image tracking. But so far, we've being using the AR Tracked Image Manager incorrectly! We populated its **Tracked Image Prefab** property with our Earth Prefab. That's a no-no. According to the Unity documentation, "`ARTrackedImageManager` has a "Tracked Image Prefab" field; however, this is not intended for content" (`https://docs.unity3d.com/Packages/com.unity.xr.arfoundation@4.2/manual/tracked-image-manager.html`). Currently, when *any* reference image is recognized, the Earth Prefab will *always* be instantiated.

Rather, when the app is in Main-mode, we should determine which planet card image is being tracked and instantiate the corresponding planet prefab for that card. So far, we only have one planet, Earth, in the image reference library. However, later in this chapter, we'll expand the project for all the planets. We can start by removing the prefab from the **AR Tracked Image Manager** component, as follows:

1. In the **Hierarchy** window, select the **AR Session Origin** game object.

2. In the **Inspector** window, delete the contents of the **AR Tracked Image Manager | Tracked Image Prefab** slot, as shown in the following screenshot:

Figure 8.16 – AR Tracked Image Manager with the default prefab field cleared

When no prefab is specified in **AR Tracked Image Manager**, an empty game object is created with an `ARTrackedImage` component on it. Now, we can instantiate the prefab as a child of that.

In our scene framework, the app starts in Startup-mode, then goes into Scan-mode once the AR Session is ready. When Scan-mode detects a reference image, it goes into Main-mode by enabling the **Main Mode** game object under **Interaction Controller**. This displays the **Main UI** panel. Let's build this panel now.

Writing the PlanetsMainMode script

In this section, we will write a new `PlanetsMainMode` script to replace the default `MainMode` one provided in the default scene template. Like other modes in our framework, it will display the appropriate UI panel when enabled. Then, when an image is tracked, it will find the corresponding planet prefab and instantiate it.

The script needs to figure out which image the AR software found and decide which prefab to instantiate as a child of the tracked image. In our case, we'll use the name of the detected image file to determine which planet card is recognized (by design, each card image is prefixed with the planet's name; for example, `Earth-MarkerCard`). The script will implement a serializable dictionary we can use to look up the planet prefab for each planet name, using the *Serialized Dictionary Lite* Asset package (you already have this package installed because `ARFramework` also requires it. See `https://assetstore.unity.com/packages/tools/utilities/serialized-dictionary-lite-110992` for more information).

Begin by performing the following steps:

1. In your **Project** `Scripts/` folder, create a new C# script named `PlanetsMainMode`.

2. In the **Hierarchy** window, select the **Main Mode** game object (under **Interaction Controller**).

3. In its **Inspector** window, remove the default **Main Mode** component using the *3-dot* context menu and selecting **Remove Component**.

4. Drag the `PlanetMainMode` script from the **Project** window onto the **Main Mode** object, adding it as a new component.

5. *Double-click* the `PlanetMainMode` script to open it for editing.

6. Begin by adding the following `using` assembly declarations at the top of the file:

```
using UnityEngine;
using RotaryHeart.Lib.SerializableDictionary;
using UnityEngine.XR.ARFoundation;
using TMPro;
using UnityEngine.UI;
```

7. When an image is tracked, we need to find which planet prefab to instantiate. At the top of the file, define a `PlanetPrefabDictionary` as follows, and declare a `planetPrefab` variable for it:

```
[System.Serializable]
public class PlanetPrefabDictionary :
SerializableDictionaryBase<string, GameObject> { }


public class PlanetsMainMode : MonoBehaviour
{
    [SerializeField] PlanetPrefabDictionary
        planetPrefabs;
```

8. When this mode is enabled, similar to the original `MainMode` script, we'll show the Main UI panel:

```
    private void OnEnable()
    {
        UIController.ShowUI("Main");
    }
```

9. Likewise, we'll enter Main-mode after Scan-mode has determined it has started tracking an image. So, `OnEnable` should also instantiate planets for the tracked images. Add a reference to `imageManager` at the top of the class:

```
    [SerializeField] ARTrackedImageManager imageManager;
```

Then, update `OnEnable`:

```
    void OnEnable()
    {
        UIController.ShowUI("Main");
        foreach (ARTrackedImage image in
                imageManager.trackables)
        {
            InstantiatePlanet(image);
        }
    }
```

This loops through the trackable images and calls `InstantiatePlanet` for each one.

10. Implement `InstantiatePlanet`, as follows:

```
void InstantiatePlanet(ARTrackedImage image)
{
    string name =
        image.referenceImage.name.Split('-')[0];
    if (image.transform.childCount == 0)
    {
        GameObject planet =
            Instantiate(planetPrefabs[name]);
        planet.transform.SetParent(image.transform,
            false);
    }
    else
    {
        Debug.Log($"{name} already instantiated");
    }
}
```

The `InstantiatePlanet` function determines the planet's name from the tracked image filename (for example, `Earth-MarkerImage`) by assuming the images follow our naming convention. It makes sure we don't already have the planet object in the scene. If not, the planet prefab is instantiated and parented to the tracked image object. (We pass `false` as a second parameter so that the planet is positioned relative to the tracked image transform. See `https://docs.unity3d.com/ScriptReference/Transform.SetParent.html`.)

11. Save the script.

12. Back in Unity, make sure you have the **Main Mode** game object selected in the **Hierarchy** window.

13. Drag the **AR Session Origin** object from the **Hierarchy** window into the **Inspector** window, dropping it onto the **Image Manager** slot.

14. In the **Inspector** window, click the + button at the bottom right of the **Planets Main Mode | Planet Prefabs** list.

15. Type the word `Earth` into the `Id` slot.

16. Unfold the item and, from the **Hierarchy** window, drag the **Earth Prefab** object on the **Value** slot in the **Inspector** window.

17. Use **File | Save** to save your work.

If you **Build and Run** now, the app will behave much the same as it did before – after Scan-mode detects an image, it enters Main-mode. But instead of AR Tracked Image Manager instantiating the Earth Prefab, instantiation is performed in `PlanetsMainMode` when it is enabled. Now, the code is ready to detect different planet card images and instantiate different corresponding planet prefabs. We will start by adding Mars.

Expanding the project with multiple planets

To add another planet to the project, we need to add its planet card image to the Reference Image Library, create its planet prefab, including a material for rendering the planet skin, and add the reference to the `planetPrefabs` list in `PlanetsMainMode`. Then, we'll update the script to handle tracking multiple planets. Let's walk through the steps for adding Mars.

Adding the planet card image to the Reference Image Library

Perform the following steps to add Mars to our **Reference Image Library**:

1. Locate and select your **ReferenceImgeLibrary** asset in the **Project** window. (If you've been following along, then it should be located in the `Image Library/` folder.)

2. In its **Inspector** window, click **Add Image**.

3. Locate and drag the `Mars-MarkerCard` image from the **Project** window and drop it onto the empty image **Texture** slot in the **Inspector** window.

4. Check the **Specify Size** checkbox and enter the same **Physical Size | X** value you used for the Earth one. Mine measures at `0.08` meters (8 cm).

5. Also, check the **Keep Texture At Runtime** checkbox.

The Reference Image Library should now look as follows:

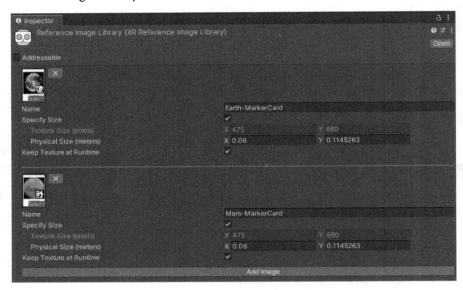

Figure 8.17 – Reference Image Library with the Mars-MarkerCard image added

Next, we'll create the Mars Prefab and material.

Creating the planet prefab

To create the planet prefab, we'll copy and modify the Earth Prefab assets. Perform the following steps:

1. In the **Project** window, locate the **Earth Prefab** asset (this will probably be in your `Prefabs/` folder).

2. Select **Edit | Duplicate** (or use *Ctrl-D* on the keyboard) to duplicate it. Rename the copy `Mars Prefab`.

3. Open **Mars Prefab** for editing. Select the child **Planet** game object.

4. In the **Project** window, *right-click* inside your `Materials/` folder and select **Create | Material**. Name it `Mars Material`.

5. Drag **Mars Material** onto the **Planet** object.

6. In the **Project** window, locate the **mars** texture file (in the `Planet Textures/` folder) and drag it onto the **Mars Material | Base Map** texture slot.

The Mars Prefab Planet should now look as follows:

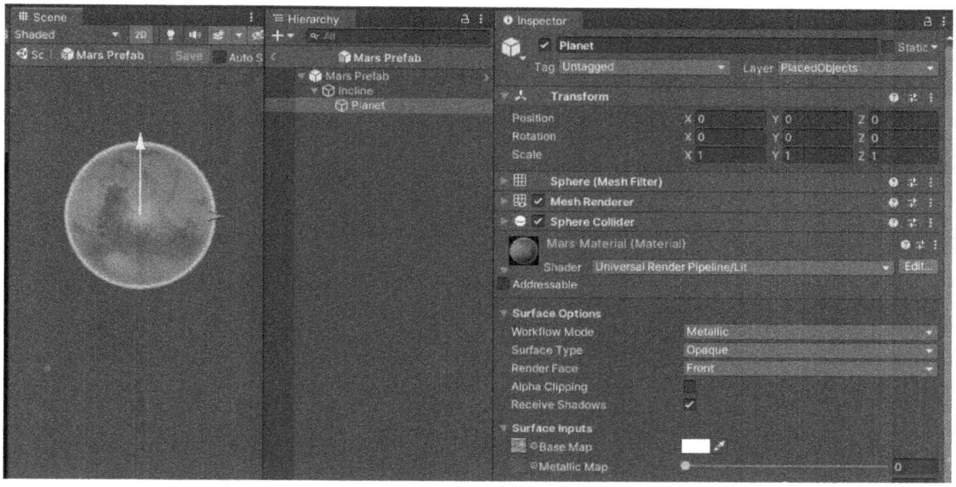

Figure 8.18 – Mars Prefab with its Planet set to the Mars Material

7. Next, we'll set the Mars Planet metadata. In the **Hierarchy** window, select the root **Mars Prefab** game object.

8. In the **Inspector** window, change the **Planet** component's parameters; that is, **Planet Name**: Mars; **Incline Degrees**: 25.2; **Rotation Period Hours**: 24.7. For its **Description**, you can add something similar to Mars is the fourth planet from the Sun and the second-smallest planet in the Solar System.

9. **Save** the prefab and return to the scene Hierarchy (using the < button at the top left of the **Hierarchy** window).

Now, we can add the prefab to the Main-mode **Planet Prefabs** dictionary, as follows:

1. In the scene **Hierarchy**, select the **Main Mode** game object (under **Interaction Controller**).

2. In the **Inspector** window, click the + button at the bottom right of the **Planets Main Mode | Planet Prefabs** list.

3. Type the word Mars into the Id slot.

4. Unfold the item and, from the **Hierarchy** window, drag the **Mars Prefab** object onto the **Value** slot in the **Inspector** window.

The **Planets Main Mode** component should now look as follows:

Figure 8.19 – The Planets Main Mode component's Planet Prefabs dictionary with Mars added

If you **Build and Run** now, when in Scan-mode, point the camera at your Mars planet card. The Mars 3D object will be added to the scene, rotating in all its glory!

Unfortunately, after doing this, if you move the camera to scan the Earth planet card, nothing will happen. Let's fix that.

Responding to detected images

Your scripts can subscribe to events so that they're notified when an image is being tracked, updated, or removed. Specifically, we can implement an OnTrackedImageChanged function to handle these events. We can use this in the PlanetsMainMode script, as follows:

1. Open the PlanetsMainMode script for editing again and add the following code:

```
void OnTrackedImageChanged
    (ARTrackedImagesChangedEventArgs eventArgs)
{
    foreach (ARTrackedImage newImage in
        eventArgs.added)
    {
        InstantiatePlanet(newImage);
    }
}
```

2. Add the following line to your `OnEnable` function, adding a listener to `imageManager`:

```
imageManager.trackedImagesChanged +=
    OnTrackedImageChanged;
```

3. Likewise, remove the listener in `OnDisable`:

```
void OnDisable()
{
    imageManager.trackedImagesChanged -=
        OnTrackedImageChanged;
}
```

When `ARTrackedImageManager` detects a new image, the Main-mode script will kick in. It contains a listener for the events and will call `InstantiatePlanet` for any newly tracked images.

4. If the app completely loses image tracking, we should go back to Scan-mode and display its instructional graphic, prompting the user to find a reference image. Add this check to `Update`, as follows:

```
void Update()
{
    if (imageManager.trackables.count == 0)
    {
        InteractionController.EnableMode("Scan");
    }
}
```

> **Tip – Tracking the State of Individual Trackables**
>
> AR Foundation also provides you with the current tracking state of each trackable image individually. Given a trackable image (`ARTRackedImage`), you can check its `trackingState` for `Tracking` – image is actively tracking, `Limited` – image is being tracked but not reliably, or `None` – the image is not being tracked. See `https://docs.unity3d.com/Packages/com.unity.xr.arfoundation@4.1/manual/tracked-image-manager.html#tracking-state`. In this project, we will only go back to Scan mode when no images are being tracked, so we don't necessarily need this extra level of status monitoring.

OK – this is getting pretty robust. **Build and Run** the project again, this time scanning either (or both) the Earth and Mars planet cards. We've got planets! The following screen capture shows the app running, with the addition of the information UI at the bottom of the screen, which we will add in the next section:

Figure 8.20 – Earth and Mars rendered at runtime

Go ahead and add the rest of the planets to your project by following these same steps. As we mentioned earlier in this chapter, referencing the NASA data provided at `https://nssdc.gsfc.nasa.gov/planetary/factsheet/index.html`, use their *Length of Day* row for our **Rotation Period Hours** parameter, and their *Obliquity to Orbit* for our **Incline Degrees** parameter. You'll notice that some planets rotate imperceptibly slowly (for example, Venus has 2,802-hour days) and spin in a direction opposite to Earth (Venus and Uranus have negative rotation periods), whereas Jupiter and Saturn rotate more than twice as fast as Earth (9.9 and 10.7 hours per day, respectively). The `Planet` script already includes an animation speed scalar, `animationHoursPerSecond`, that you can use to modify the rotation rates that are visualized in the app.

Now that our application supports multiple planets, you might want to tell the user more about the specific planet that they are looking at. Let's add this capability to Main-mode so that it responsively updates the UI.

Making a responsive UI

In this section, we'll add an info panel to the bottom of the screen (as shown in the preceding screen capture). When you point the camera at one planet or another, we'll show the planet's name, as well as an **Info** button, which will cause a text box to appear that contains more information about that planet.

Creating the Main-mode UI

When the app is in Main-mode, the Main UI panel is displayed. On this panel, we'll show the name of the current planet and an **Info** button for the user to press when they want more details about that planet. Perform the following steps:

1. In the **Hierarchy** window, unfold the **UI Canvas** object and unfold its child **Main UI** object.

2. The default child text in the panel is a temporary placeholder, so we can remove it. *Right-click* the child **Text** object and select *Delete*.

3. Create a subpanel by *right-clicking* on **Main UI** and selecting **UI | Panel**. Rename it `Info Panel`.

4. Use **Anchor Presets** to set **Bottom-Stretch**. Then, use *Shift + Alt* + **Bottom-Stretch** to make a bottom panel. Then, set its **Rect Transform | Height** to `175`.

5. I set my background **Image | Color** to opaque white with **Alpha:** `255`.

6. Create a text element for the planet name. *Right-click* **Info Panel** and select **UI | Text – TextMeshPro**. Rename the object `Planet Name Text`.

7. On the Planet Name Text **TextMeshPro – Text** component's **Text Input**, set the content with a temporary string such as `[Planet name]`.

8. Set the text properties; for example, **Anchor Presets: Stretch-Stretch** (and *Shift + Alt* + **Stretch-Stretch**); **Text Align: Left, Middle**; **Rect Transform | Left:** `50`; **Text Vertex Color:** black; **Font Size:** `72`.

9. Create an **Info** button. *Right-click* **Info Panel** and select **UI | Button – TextMeshPro**. Rename it `Info Button`.

10. Set the button properties; for example, **Anchor Presets: Right-Middle** (and *Shift + Alt* + **Right-Middle**); **Width** and **Height:** `150, 150`; **Pos X:** `-20`.

11. Unfold **Info Button**. On its child **Text** object, set **Top:** `-50` and its text content to `Info`.

12. *Right-click* **Info Button** and select **UI | Text – TextMeshPro**. On the new text element, set its text value to a question mark, `?`, **Font Size** to `72`, **Color** to black, **Alignment** to **Center, Middle**, and **Pos Y** to `-15`.

13. We're going to use this button to toggle a details panel on and off. So, let's replace its **Button** component with a **Toggle** instead. With the **Info Button** object selected in the **Hierarchy** window, in the **Inspector** window, remove the **Button** component using the *3-dot* context menu and selecting **Remove Component**.

14. Select **Add Component**, search for `toggle`, and add a **Toggle** component.

My main **Info Panel** now looks as follows:

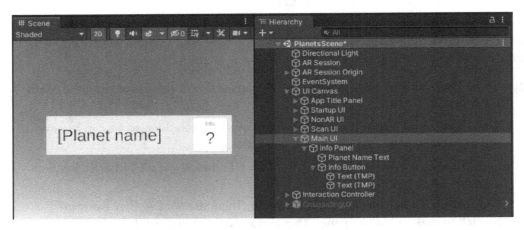

Figure 8.21 – Main UI's Info Panel with Planet Name Text and an Info Button

The Planet Name Text's content will be filled in at runtime. Let's add that code now.

Pointing the camera to show information

The plan is that with one or more virtual planets instantiated in the scene, the user can point the camera at a planet so that it displays the planet's name in the Info Panel. This can be implemented using a `Physics Raycast`. (Raycasts were introduced and explained in the previous chapters. See `https://docs.unity3d.com/ScriptReference/Physics.Raycast.html`). Recall that at the beginning of this chapter, we put the Planet Prefab on a layer named `PlacedObjects`. We'll make use of that here.

Make the following changes to the `PlanetsMainMode` script:

1. Ensure the script file contains the following assembly references:

```
using TMPro;
using UnityEngine.UI;
```

2. At the top of the class, declare and initialize references to the AR `camera` and `layerMask` variables, as follows:

```
Camera;
int layerMask;

void Start()
{
    camera = Camera.main;
    layerMask =
        1 << LayerMask.NameToLayer("PlacedObjects");
}
```

3. Also, add references to the `planetName` and `infoButton` UI elements in the Info Panel:

```
[SerializeField] TMP_Text planetName;
[SerializeField] Toggle infoButton;
```

4. We can initialize the UI settings when the mode is enabled. Please add the following lines to the `OnEnable` function:

```
planetName.text = "";
infoButton.interactable = false;
```

5. Then, add the following highlighted code to the `Update` function:

```
void Update()
{
    if (imageManager.trackables.count == 0)
    {
        InteractionController.EnableMode("Scan");
    }
    else
    {
        Ray = new Ray(camera.transform.position,
            camera.transform.forward);
        RaycastHit hit;
        if (Physics.Raycast(ray, out hit,
            Mathf.Infinity, layerMask))
```

```
        {
            Planet = hit.collider.
                GetComponentInParent<Planet>();
            planetName.text = planet.planetName;
            infoButton.interactable = true;
        }
        else
        {
            planetName.text = "";
            infoButton.interactable = false;
        }
    }
}
```

6. Save the script. Back in Unity, select the **Main Mode** object in the **Hierarchy** window.

7. Drag the **Planet Name Text** game object from the **Hierarchy** window (under **UI Canvas / Main UI / Info Panel**) into the **Planets Main Mode | Planet Name** slot.

8. Drag the **Info Button** object onto the **Info Button** slot.

Go ahead and **Build and Run** the project one more time. While viewing one or more planets, as you point the device's camera at one of them, the planet's name will be shown in the Info Panel at the bottom of the screen.

Lastly, let's set up the Info Button and description display.

Displaying information details

When the user is pointing their camera at a virtual 3D planet in the scene, we show the name of the planet in the Info Panel. When the user clicks the **Info** button, we want to show more information about the planet, such as its description text. Let's add a text panel for that now by performing the following steps:

1. In the **Hierarchy** window, select the **Main UI** game object (under **UI Canvas**) and *right-click* and select **UI | Panel**. Rename it `Details Panel`.

2. It has already been set to **Stretch-Stretch**, which is what we want. But let's adjust its size. Set **Left**: `30`, **Right**: `30`, **Top**: `150`, and **Bottom**: `200`.

3. *Right-click* **Details Panel** and select **UI | Text – TextMeshPro**. Rename it `Details Text`.

4. Format the text area; for example, set its **Anchor Presets** to **Stretch-Stretch** (and *Shift + Alt* + **Stretch-Stretch**), its **Text Vertex Color** to black, its **Font Size** to 48, its **Rect Transform Left, Right, Top, Bottom** to 30, 30, 30, 30, and its **Alignment**: to **Center, Middle**.

Now, add control of this panel to the `PlanetsMainMode` script, as follows:

1. Add references to `detailsPanel` and `detailsText` at the top of the class:

```
[SerializeField] GameObject detailsPanel;
[SerializeField] TMP_Text detailsText;
```

2. Ensure the panel is hidden when the mode is enabled. Add the following line to the `OnEnable` function:

```
detailsPanel.SetActive(false);
```

3. Initialize the panel's content when a planet is being selected. That is, in `Update`, we must set `detailsText` at the same time we set `planetName`:

```
if (Physics.Raycast(ray, out hit,
    Mathf.Infinity, layerMask))
{
    Planet = hit.collider.
        GetComponentInParent<Planet>();
    planetName.text = planet.planetName;
    detailsText.text = planet.description;
    infoButton.interactable = true;
}
else
{
    planetName.text = "";
    detailsText.text = "";
    infoButton.interactable = false;
}
```

Save the script. Back in Unity, we'll wire up the **Info Button** toggle.

4. With **Info Button** selected in the **Hierarchy** window, in the **Inspector** window, click the + button at the bottom right of the **Toggle | On Value Changed** action list.

5. From the **Hierarchy** window, drag the **Details Panel** game object onto the **On Value Changed | Object** slot.

6.   From the **Function** selection list, choose **GameObject | Dynamic Bool | SetActive**.

7.   Save the scene.

Now, when you **Build and Run** the project and view a planet, then press the **Info** button, the Details Panel will be displayed alongside the planet's description text, as shown in the following screen capture from my phone:

Figure 8.22 – Displaying description text about Mars in the toggled Details Panel

In this section, we added a responsive UI to the scene. When the user points their device camera at a virtual planet that's been instantiated in the scene, the name of the planet is displayed in the Info Panel at the bottom of the screen. If the user taps the **Info** button, a text panel is toggled, showing additional details about that specific planet.

Can you think of additional ways to improve this project?

Summary

In this chapter, you built an AR project that lets you visualize and learn about planets in our Solar System. The scene uses AR image detection and tracks the planet cards that you printed out from the PDF file provided with the files for this book. Each planet card image includes a distinct marker with unique details, high contrast edges, and asymmetric shapes, making them readily detectable and trackable by the AR system. You set up the AR Session to track images using the AR Trackable Image Manager component and built a Reference Image Library asset with the planet card images.

You then created a generic Planet Prefab with a Planet script that controls the rotation behavior and metadata for a planet. Then, you created separate prefab variants for each planet. You wrote a `PlanetsMainMode` script that instantiates the correct planet prefab when a specific planet card image is detected. This allows multiple tracked images and planets to be present in the scene. Then, you added a responsive UI where the user can point their device camera to an instantiated planet and get additional information about that virtual object.

In the next chapter, we'll explore another kind of AR application: flipping the device camera so that it's facing the user to make selfie face filters.

9
Selfies: Making Funny Faces

In this chapter, you will learn how to use Unity AR Foundation for face tracking in order to make fun and entertaining face filters. I apologize in advance for showing my handsome face throughout this chapter – it's a necessary evil when working with selfies!

We'll start with a brief explanation of how face tracking works, and then we will create a new AR scene with face tracking enabled. We will use a couple of 3D head models that track your head pose and to which you can add extra accessories, such as a hat and sunglasses. We are going to build a main menu so that the user can select and change models at runtime. We'll then work with dynamic face meshes and create several materials to easily switch between them. In the last part, we'll look at more advanced features such as eye tracking, face regions (ARCore), and blend shapes (ARKit).

We will cover the following topics:

- Understanding face tracking
- Configuring a new AR scene for face tracking
- Tracking the face pose with 3D models and accessories
- Controlling the app's main mode and building a main menu
- Making dynamic face meshes with a variety of materials

- Using eye-tracking (ARKit)
- Attaching stickers to face regions (ARCore)
- Tracking expressive face blend shapes (ARKit)

By the end of this chapter, you'll be familiar with many of the face tracking features in AR Foundation, ARCore, and ARKit. You will also have a working *Face Maker* project you can show off to your friends!

Technical requirements

To implement the project in this chapter, you need Unity installed on your development computer and connected with a mobile device that supports augmented reality applications (see *Chapter 1, Setting Up for AR Development*, for instructions). We also assume that you have the `ARFramework` template and its prerequisites installed (see *Chapter 5, Using the AR User Framework*). The completed project can be found in this book's GitHub repository, available at the following URL: `https://github.com/PacktPublishing/Augmented-Reality-with-Unity-AR-Foundation`.

Understanding face tracking

Let's start with some background on face tracking and the technology that makes it work. Face tracking is a kind of Augmented Reality that (usually) uses the front-facing camera on your mobile device. Apps such as Snapchat, Instagram, and Animoji have popularized face filter technology, and it has now become mainstream on mobile devices. It makes for highly entertaining and creative experiences. The technology detects facial features and expressions, and Unity AR Foundation enables you to write applications for attaching 3D objects to specific facial features that are tracked.

Face tracking begins with a frame of the video from your device's camera. It analyzes the pixels, looking for patterns that represent a face – for example, the bridge of the nose is lighter than the pixels surrounding it, and the eyes are darker than the forehead. Key points and regions are recognized and used to construct a 3D mesh, like a mask, representing the face. Nodes of the mesh are "locked onto" key points in the image, allowing the mesh to follow not just the pose of the face, but detailed changes that correspond to human facial expressions, like a smile or a wink of the eye.

To learn more about how face tracking works, I encourage you to watch the seminal Vox video (over 3 million views) *How Snapchat's filters work*, available at the following URL: `https://www.youtube.com/watch?v=Pc2aJxnmzh0`.

It's helpful to understand the distinction between face tracking and face identification, and how to track a face with AR Foundation.

Face tracking versus face identification

A distinction should be made between *face tracking* and *face identification*. Face tracking, in general, is limited to detecting a human face and tracking its pose (position and rotation), facial features such as the forehead and nose, and changes representing expressions, such as opening your mouth or blinking your eyes. Face identification, on the other hand, adds recognition of the features that make your face unique and different from other faces. Face recognition is used as a fingerprint. One example of face recognition technology is for unlocking devices. More advanced (and creepy) face identification is increasingly being used by authoritarian governments and law enforcement to identify strangers in a crowd, using a large database of faces.

Using Unity AR Foundation, you can access the AR face-tracking capabilities of your device. We are going to examine this next.

Tracking a face with AR Foundation

As you now know, a Unity project using AR Foundation and XR Plugins will have a scene that includes an **ARSession** and an **ARSessionOrigin** object. The AR Face Manager component is added to the AR Session Origin to enable face tracking. Like most AR Foundation features, this component wraps the Unity AR subsystems, namely the XR face subsystem (see `https://docs.unity3d.com/Packages/com.unity.xr.arsubsystems@4.2/api/UnityEngine.XR.ARSubsystems.XRFaceSubsystem.html`). This in turn interfaces with the underlying XR plugin, such as ARCore or ARKit.

The **AR Face Manager** component references a face prefab provided by you. This prefab will be instantiated and tracked with the detected face. The component also provides a **Maximum Face Count** parameter, should you want the app to support multiple people in the same camera view (depending on the capabilities of the underlying device). The component is shown in the following screenshot:

Figure 9.1 – The AR Face Manager component on an AR Session Origin object

The face prefab should have an **AR Face** component on it that represents a face detected by an AR device. It has properties including the face mesh vertices, facet normals, and transforms for the left and right eyes. Like other AR trackables, your scripts can subscribe to changes to know when faces have been added, updated, and removed. The specific properties available will depend on the capabilities of the underlying device. See the documentation available at the following URL: `https://docs.unity3d.com/Packages/com.unity.xr.arfoundation@4.2/api/UnityEngine.XR.ARFoundation.ARFace.html`. Also, see the following URL: `https://docs.unity3d.com/Packages/com.unity.xr.arsubsystems@4.2/api/UnityEngine.XR.ARSubsystems.XRFace.html`.

AR Foundation provides an interface for AR face tracking (not identification), using the AR Face Manager component added to your AR Session Origin object. We can now get started building a selfie face filter project.

Getting started

To begin, we'll create a new scene named `FaceMaker` using the `ARFramework` scene template. If you're targeting iOS ARKit, there may be additional setup required, including installing the separate ARKit Face Tracking package. Then we'll add a project title to the UI before moving on to adding face tracking to the scene.

Creating a new scene using the ARFramework template

Create a new scene in your Unity AR-ready project using the following steps:

1. Select **File | New Scene**.
2. In the **New Scene** dialog box, select the **ARFramework** template.
3. Click **Create**.
4. Select **File | Save As**. Navigate to the `Scenes/` folder in your project `Assets` folder, give it the name `FaceMaker`, and click **Save**.

The new AR scene already has the following setup from the template:

*   **AR Session** game object with an AR Session component.
*   **An AR Session Origin** rig with an AR Session Origin component, among others, and a child main camera. We will replace its AR Plane Manager component with an AR Face Manager one.

- **UI Canvas** is a screen space canvas with the child panels **Startup UI**, **Scan UI**, **Main UI**, and **NonAR UI** that we built for the **ARFramework**. It has the UI Controller component script that we wrote. We'll update this with the project-specific UI.

- **Interaction Controller** is a game object we built for the ARFramework, with an interaction controller component script we wrote that helps the app switch between interaction modes, including Startup, Scan, Main, and NonAR modes. It also has a **Player Input** component configured with the **AR Input Actions** asset we previously created. We are going to customize the main mode for our face tracking app.

- **OnboardingUX** is a prefab from the AR Foundation Demos project that provides AR session status messages and animated onboarding graphics prompts.

Let's start by setting the app title now as follows:

1. In the **Hierarchy**, unfold the **UI Canvas** object, and unfold its child **App Title Panel**.
2. Select the **Title Text** object.
3. In its **Inspector**, change its text content to `Face Maker`.

If you are targeting ARKit on iOS, there may be additional project setup required.

Setting up iOS ARKit for face tracking

To develop and build a project using face tracking with ARKit for an iOS device, you also need to install the ARKit Face Tracking package via the package manager. Perform the following steps:

1. Open the package manager using **Window | Package Manager**.
2. In the **Packages** filter selection at the top left, choose **Unity Registry**.
3. Use the search field at the top right to search for `ar`, and select the **ARKit Face Tracking** package from the packages list.
4. Click **Install** at the bottom right of the window.

Then, configure ARKit XR Plugin for face tracking, as follows:

1. Open the **Project Settings** window, using **Edit | Project Settings**.
2. On the left-side tabs menu, select **XR Plug-in Management | ARKit**.
3. Check the **Face Tracking** checkbox.

Next, we will gather some assets that we'll be using in this chapter. Some of these are also provided in this book's GitHub repository. Others are third-party assets that you must download and import separately.

Importing assets used in this project

First, you should already have the *AR Foundation Samples* assets in your project (the ones that we imported back in *Chapter 1, Setting Up for AR Development*). If you followed along, these are in the `Assets/ARF-samples/` folder. It contains some useful example assets that we'll use and reference in this chapter that can give you additional insight into the capabilities of AR Foundation face tracking, as well as how to use those capabilities.

We are also going to use the assets from the *AR Face Assets* package from Unity (available in the Asset Store). These assets are also used in the Unity Learn tutorial, *AR Face Tracking with AR Foundation* (`https://learn.unity.com/project/ar-face-tracking-with-ar-foundations`). To import the package, follow these steps:

1. Using your internet browser, go to the following URL: `https://assetstore.unity.com/packages/essentials/asset-packs/ar-face-assets-184187`.

2. Click **Add to My Assets** (if necessary), then click **Open In Unity**.

3. In Unity, this should open the **Package Manager** window (or select **Window | Package Manager**).

4. Select **My Assets** from the **Packages** filter at the top left.

5. Find the **AR Face Assets** package and click **Download** and/or **Import** (bottom right). In the **Import Unity Package** window, click the **Import** button.

6. Convert the imported materials to the Universal Render Pipeline by selecting **Edit | Render Pipeline | Universal Render Pipeline | Upgrade Project Materials to URP Materials**.

Face accessories 3D models: I have found some free 3D models to use in this project. You can also use them or substitute your own. If you wish to use them, they are included in the following GitHub repositories:

- Sunglasses: `https://free3d.com/3d-model/sunglasses-v1--803862.html`. OBJ format (submitted by *printable_models*, `https://free3d.com/user/printable_models`)

- Top hat: `https://free3d.com/3d-model/cartola-278168.html`. FBX format (submitted by *zotgames*, `https://free3d.com/user/zotgames`)

If you're downloading these yourself, unzip and drag the files into your project's `Assets/` folder. We'll address the import settings and steps later in the chapter.

Face stickers 2D sprite images: For the ARCore-based face region stickers, I found some free clipart at Creative Commons. You can use them or substitute your own. If you wish to use them, they are included in the following GitHub repositories:

- Eyebrows: `https://clipground.com/images/angry-eyebrows-clipart-11.png`

- Mustache: `https://clipground.com/images/monocle-clipart-12.jpg`

- Licking lips: `https://clipground.com/images/licking-lips-clipart-12.jpg`

I used Photoshop to adapt each of these images with a transparent background, square-shaped canvas, and scaled to 512x512 pixels. These are imported as **Texture Type: Sprite (2D and UI)**.

For all the aforementioned assets, I also created button icons that we'll use in the UI. These are also available on the GitHub repository in the `icons/` folder and are imported as **Texture Type: Sprite (2D and UI)**.

We now have our basic scene created, as well as prerequisite assets imported into the project. We used the `ARFramework` scene template created for this book when creating the new scene, and updated the UI title text for this project. If you're working on iOS, we also installed extra required packages into the project. Then, we imported other graphic assets we're going to use, including the demo AR Face Assets pack provided by Unity. Let's now configure the scene for face tracking.

Configuring a new AR scene for face tracking

There are a few simple steps required to configure an AR Foundation-based scene for face tracking. Since we're going to do selfies, we'll set up the AR camera to use input from the front-facing camera. Then we'll add an AR Face Manager component to the AR Session Origin. If you want to use the Unity Onboarding UX animated graphic to prompt the user, you can adapt the `ScanMode` script for that.

Let's get started!

Setting the AR camera for selfies

Use the following steps to set up the AR camera for selfies:

1. In the **Hierarchy**, unfold the **AR Session Origin** game object and select its child **Main Camera**.

2. In the **Inspector**, set **AR Camera Manager | Facing Direction** to **User**.

3. We also need to set the AR Session tracking mode for rotation only. Select the **AR Session** game object in the **Hierarchy**.

4. In the **Inspector**, set the **AR Session | Tracking Mode** to **Rotation Only**.

Next, we'll add the AR Face Manager component to the AR Session Origin.

Adding an AR Face Manager component

Using the scene provided by the `ARFramework` template, we will replace the given AR trackable components with an **AR Face Manager** one. For the **Face Prefab**, we'll start with the **TriAxes** prefab from the AR Samples project. If you examine this prefab, you'll discover it has an **AR Face** component, so it can be used as a trackable.

To configure the **AR Session** to track faces, follow these steps:

1. In the **Hierarchy** window, select the **AR Session Origin** game object.

2. In the **Inspector** window, use the *three-dot context menu* (or *right-click*) on the **AR Plane Manager** component, and select **Remove Component**.

3. Use the **Add Component** button, search for `AR`, and add an **AR Face Manager** component.

4. In your **Project** window, locate the **TriAxes** prefab file (likely in the `Assets/ARF-samples/Prefabs/` folder), and drag it into the **Inspector**, dropping it onto the **AR Face Manager | Face Prefab** slot.

5. Save the scene using **File | Save**.

The scene is now basically set up for face tracking. `ARFamework` includes a scan mode that prompts the user to find a trackable object with their camera. We can now configure that for face tracking.

Prompting the user to find a face, or not

Optionally, you can have your app prompt the user to scan for their face. This is less necessary when using the selfie camera (with **Facing Direction** set to **User**) because when you're holding your phone and looking at the screen, the camera is looking right back at you! But if your app were using the world-facing camera instead, it might be necessary to use an instructional prompt to tell the user to find a face.

To skip the scan mode and its instructional prompt, tell the startup mode to proceed directly to the main mode, using the following steps:

1. In the **Hierarchy**, under the **Interaction Controller** game object, select the **Startup Mode** object.

2. In the **Inspector**, enter the text Main into the **Next Mode** property.

Otherwise, if you want to use scan mode, you'll need to write a FaceScanMode script as follows:

1. In your **Project** window's Scripts/ folder, *right-click* and select **Create | C# Script**. Name it FaceScanMode.

2. Open the script for editing and replace its contents as follows:

```
using UnityEngine;
using UnityEngine.XR.ARFoundation;

public class FaceScanMode : MonoBehaviour
{
    [SerializeField] ARFaceManager faceManager;

    private void OnEnable()
    {
        UIController.ShowUI("Scan");
    }

    void Update()
    {
        if (faceManager.trackables.count > 0)
        {
```

```
        InteractionController.EnableMode("Main");
    }
  }
}
```

The script shows the **Scan UI** panel and then, in `Update`, waits until a face is being tracked before transitioning the app to the main mode.

3. In Unity, in the **Hierarchy** window, select the **Scan Mode** object (under **Interaction Controller**).

4. Remove the old **Scan Mode** component using the three-dot context menu and then choose **Remove Component**.

5. Drag the new `FaceScanMode` script onto the **Scan Mode** game object, adding it as a component.

6. Drag the **AR Session Origin** game object from the **Hierarchy** onto the **Face Scan Mode | Face Manager** slot.

7. In the **Hierarchy**, navigate and select **UI Canvas | Scan UI | Animated Prompt**.

8. In the **Inspector**, set the **Instruction** property to **Find A Face**.

With this latter setup, the app starts in startup mode. After the AR Session is running, it goes to scan mode, prompting the user to find a face. Once a face is detected, the app proceeds to main mode (as yet, this does nothing). You also have the option to skip the scan mode prompt altogether by telling the startup mode to go straight to the main mode.

Let's make sure everything works so far. You're now ready to try to run the scene.

Build and run

Let's do a **Build And Run** on your device to ensure the project is set up correctly. Use the following steps:

1. Save your work using **File | Save**.

2. Select **File | Build Settings** to open the **Build Settings** window.

3. Click **Add Open Scenes** to add the `FaceMaker` scene to **Scenes In Build**, and ensure it is the only scene in the list with a checkmark.

4. Ensure your target device is plugged into a USB port and is ready.

5. Click **Build And Run** to build the project.

In the following screen capture, you can see the face pose is visualized using the **TriAxes** prefab. I have tilted my head to the side and back a little to make the three axes more evident.

Figure 9.2 – Tracking the face pose, visualized with the TriAxes prefab

Note the direction of each of the axes. The axes are colored red, green, and blue, corresponding to X, Y, and Z respectively. The positive Z direction is in the direction that the device camera is facing, and thus, pointing towards my back.

Now that we have face tracking running, let's substitute this **TriAxes** prefab with something more interesting – a whole 3D head model.

Tracking the face pose with 3D heads

The AR Face Assets package from Unity that we imported at the top of this chapter contains a couple of 3D head models we can use in our project. We'll create prefabs of each model and try them separately in the AR Face Manager **Face Prefab** property. In the next section, we'll create a menu so that the user can pick which head to view at runtime.

Making a Mr. Plastic Head prefab

The first head prefab will use the Plasticscene Head assets given in the Unity AR Face Assets package, and found in the `Assets/AR face Assets/3D Head/ Plasticene Head/` folder. This folder contains an FBX model named `Plasto_Head` and a material named `PlasiceneHead` (the typo is theirs). The model will require some transform adjustments before it can be used as a face prefab. To create a prefab for this model, use the following steps:

1. In the **Project** window, *right-click* in your `Prefabs/` folder (create one first if necessary) and choose **Create | Prefab**. Rename it `MrPlasticHead`.

2. Click **Open Prefab** to begin editing.

3. With the root object selected, in the **Inspector**, click **Add Component**. Then, search and choose **AR Face** to add an **AR Face** component.

4. From the **Project** window, drag the `Plastic_Head` model to the **Hierarchy** and drop it as a child of **MrPlasticHead.**

5. Select the **Plasto_Head** object in the **Hierarchy**. In the **Inspector**, set **Rotation | Y** to `180`, so it's facing the camera.

6. Set **Scale** to (`0.6, 0.6, 0.6`). Then set **Position | Y** to `-0.2`. I selected these transform settings by trial and error and using a measuring cube (see the inset *Tip*).

7. If the default material (converted to URP) appears too dark, select the child **Plaso_Head/Plasto_Head** object, and in the **Inspector**, under the **Plasicene Head** material, set the **Base Map** color to white (from middle gray).

8. Save the prefab and exit back to the scene **Hierarchy** window using the < button in the top left of the window.

> **Tip: Measuring faces using a cube object**
>
> A human head is approximately 0.125 meters (5 inches) wide. You can use this fact for scaling 3D models used in your face prefabs. To help judge this, while editing a face prefab, try adding a 3D cube object, set **Position** to (`0, 0, 0`), **Rotation** to (`0, 0, 0`), and **Scale** to (`0.125, 0.125, 0.125`). This can help you decide the transform parameters of other imported models you are using.

Let's see how this looks. Add the prefab to the **AR Face Manager** and build the project as follows:

1. In the **Hierarchy** window, select the **AR Session Origin** game object.

2. From the **Project** window, drag the **MrPlasticHead** prefab into the **Inspector**, dropping it onto **AR Face Manager | the Face Prefab** slot.

3. Save the scene using **File | Save**.

4. Build the project using **File | Build And Run**.

Info: Material texture maps

You may note that the `PlasticeneHead` material uses three textures for the **Base** (albedo), **Normal**, and **Occlusion** maps. The **Base** texture provides the albedo coloring as if the surface of the mesh were painted with these pixels. The Normal map (also known as the Bump map or Height map) lets the shader alter the mathematical surface normal vector in more detail than given by the mesh geometry itself, simulating surface textures that are especially noticeable with lighting. Finally, the **Occlusion** map provides additional realism by darkening deeper crevasses in the surface texture, creating higher contrast as occurs in real-life materials. For a more detailed explanation, starting with **Normal** maps, see the following URL: `https://docs.unity3d.com/Manual/StandardShaderMaterialParameterNormalMap.html`.

A screen capture of me with a Mr. Plastic Head head is shown below, together with the Mr. Facet Head model that we'll use next:

Figure 9.3 – Screen capture of myself with MrPlasticHead (right) and MrFacetHead (left)

Let's make the **MrFacetHead** prefab next.

Making a Mr. Facet Head prefab

There is a second model provided in the AR Face Assets package, Faceted Head, found in the `Assets/AR face Assets/3D Head/Faceted Head/` folder. This folder contains an FBX model named `FacetedHead`, and a material also named `FacetedHead`. As before, the model will require some transform adjustments to be used as a face prefab. To create a prefab for this model, use the following steps:

1. In the **Project** window, *right-click* in your `Prefabs/` folder and choose **Create | Prefab**. Rename it `MrFacetHead`.

2. Click **Open Prefab** to begin editing.

3. With the root object selected, in the **Inspector**, click **Add Component**. Then, search and choose **AR Face** to add an **AR Face** component.

4. From the **Project** window, drag the `FacetedHead` model to the **Hierarchy** and drop it as a child of **MrFacetHead**.

5. With the **FacetedHead** object selected in the **Hierarchy**, in the **Inspector**, set **Rotation | X** to `-90` so that it's facing the camera. Set **Scale** to (`1.1, 1.1, 1.1`).

6. If the default material (converted to URP) appears too dark, select the **FacetedHead** object, and in its **Inspector** under the **FacetedHead** material, set the **Base Map** color to white.

7. Save the prefab, and exit back to the scene **Hierarchy** window, using the < button at the top left of the window.

8. In the **Hierarchy** window, select the **AR Session Origin** game object.

9. From the **Project** window, drag the **MrFacetHead** prefab into the **Inspector**, dropping it onto the **AR Face Manager | Face Prefab** slot.

10. Save the scene using **File | Save**.

11. Build the project using **File | Build And Run**.

When it runs, you now have a Mr. Faceted Head head, as shown in the preceding figure (yes, those are my real eyes peering through the mask).

In this section, we created two prefabs, **MrPlasticHead** and **MrFacetHead**, using assets from the Unity *AR Face Assets* package that we imported earlier. Each of these has an AR Foundation **AR Face** component on its root GameObject and different imported models for the two heads. We tried using one of these in our app by adding it to the **AR Face Manager** component and running the scene.

Wouldn't it be nice to let the user choose a head at runtime, rather than manually setting the AR Face Manager and rebuilding the project? Next, let's create a main menu, and a changeable face prefab we can control from the menu buttons.

Building the Main mode and menu

In this section, we will set up the main mode app to handle user interactions, including face filter selections from a main menu. To do this, we first need to create a changeable face prefab that can be told which facial features to display. We'll write a `FaceMainMode` script that displays the main UI panel and passes change requests from the user to the face object. Then, we'll make a main menu with a set of horizontally scrolling buttons that the user can tap to change face filters.

Creating a changeable face prefab

To create a face prefab that we can use for dynamically changing filters during runtime, we'll start with an empty game object with an AR Face component, and add a script for setting the contained prefab object. Use the following steps:

1. In the **Project** window, *right-click* in your `Prefabs/` folder and choose **Create |
    Prefab**. Rename it `Changeable Face Prefab`.

2. Click **Open Prefab** to begin editing.

3. With the root object selected, in the **Inspector**, click **Add Component**. Search for and choose **AR Face** to add an **AR Face** component.

4. In your **Project** window's `Scripts/` folder, *right-click* and select **Create |
    C# Script**, then name it `ChangeableFace`.

5. Open the script for editing and replace its contents as follows:

```
using System.Collections.Generic;
using UnityEngine;
using UnityEngine.XR.ARFoundation;

public class ChangeableFace : MonoBehaviour
{
    GameObject currentPosePrefab;
    GameObject poseObj;

    public void SetPosePrefab(GameObject prefab)
    {
        if (prefab == currentPosePrefab)
            return;

        if (poseObj != null)
```

```
            Destroy(poseObj);

        currentPosePrefab = prefab;
        if (prefab != null)
            poseObj = Instantiate(prefab, transform,
                false);
    }
}
```

The script exposes a public `SetPosePrefab` function that instantiates the `prefab` argument as a child of the current object. If the requested prefab is already instantiated, the request is ignored. If there was a previously instantiated object, it is first destroyed. The function can be called with a null value for the `prefab` argument that will only clear the existing instantiated object.

6. Save the script and, back in Unity, drag the `ChangeableFace` script onto the **Changeable Face Prefab** root object.

7. Save the prefab and exit back to the scene hierarchy using the < button at the top left of the window.

8. In the **Hierarchy**, select the **AR Session Origin** object. From the **Project** window, drag the **Changeable Face Prefab** into the **Inspector**, dropping it onto the **AR Face Manager | Face Prefab** slot.

We now have a **Changeable Face Prefab** asset with a `ChangeableFace` script. We are planning to call its `SetPosePrefab` function from the main mode in response to a user button click. We should set up the main mode now.

Writing a main mode controller script

In our ARFramework template, interaction modes are represented with game objects under the interaction controller and are activated when a specific mode is enabled. The default `MainMode` script from the template is simply a placeholder. We should replace it now with a custom script for this project. To do so, follow these steps:

1. In the **Project** window, *right-click* and select **Create | C# script**, naming it `FaceMainMode`.

2. In the **Hierarchy**, select the **Main Mode** game object (under **Interaction Controller**).

3. In the **Inspector**, remove the default **Main Mode** component using the three-dot menu, then click **Remove Component**.

4. Drag the new `FaceMainMode` script onto the **Main Mode** object, adding it as a component.

5. Open the `FaceMainMode` script for editing, and start it as follows:

```
using UnityEngine;
using UnityEngine.XR.ARFoundation;

public class FaceMainMode : MonoBehaviour
{
    [SerializeField] ARFaceManager faceManager;

    void OnEnable()
    {
        UIController.ShowUI("Main");
    }

    public void ChangePosePrefab(GameObject prefab)
    {
        foreach (ARFace face in faceManager.trackables)
        {
            ChangeableFace changeable =
                face.GetComponent<ChangeableFace>();
            if (changeable != null)
            {
                changeable.SetPosePrefab(prefab);
            }
        }
    }
}
```

When the main mode is enabled, it shows the main UI panel. This will contain the main menu buttons. When a menu button is clicked and it calls `ChangePosePrefab`, that in turn will call `SetPosePrefab` for any trackable faces in the scene.

Let's create the menu UI next.

Creating scrollable main menu buttons

In our user framework, a mode's UI panel will be enabled by the corresponding interaction mode. We'll now add a horizontally-scrolling main menu to the main UI panel with buttons that can change the tracked face. Use the following steps:

1. In the **Hierarchy**, select the **Main UI** game object (under **UI Canvas**), then *right-click* and select **UI | Panel**. Rename the new panel to `MainMenu Panel`.

2. In the **Inspector**, use the **Anchor Presets** option (at the upper left of **Rect Transform**) to set **Bottom-Stretch**, then use *Shift + Alt + left-click Bottom-Stretch*.

3. Set **Rect Transform | Height** to `150`.

4. Remove its **Image** component with the three-dot menu, then **Remove Component** (we won't have a background on this menu).

5. In the **Hierarchy**, *right-click* the **MainMenu Panel**, and select **UI | Scroll View**.

6. Use **Anchor Presets** to click the **Stretch-Stretch** option, and then use *Shift + Alt + left-click Stretch-Stretch*.

7. Remove the **Image** component.

8. In the **Scroll Rect** component, uncheck **Vertical**.

9. Delete the content of the **Horizontal Scrollbar** and **Vertical Scrollbar** fields, and disable (or delete) the **Scrollbar Horizontal** and **Scrollbar Vertical** game objects in the hierarchy.

10. In the **Hierarchy**, unfold the child **Viewport** game object, and select the child **Content** game object.

11. Click **Add Component**, then search for and select **Horizontal Layout Group**.

12. Uncheck all of its checkboxes, including **Child Force Expand | Width** and **Height**.

13. Set **Spacing** to `5`.

14. Click **Add Component**, then search for and select **Content Size Fitter**.

15. Set **Horizontal Fit** to **Preferred Size**.

We now have a **MainMenu Panel** under **Main UI**. It contains a horizontally-scrolling content area, as shown in the following screenshot of the UI hierarchy with the **Content** object selected:

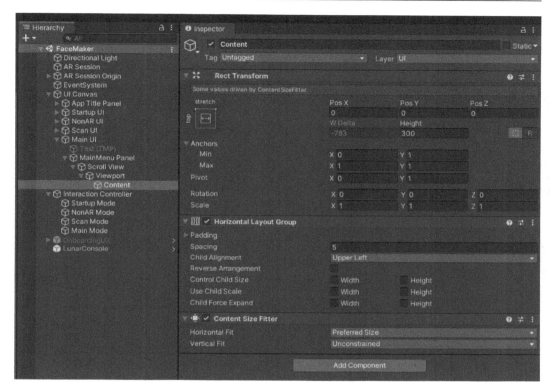

Figure 9.4 – Main UI hierarchy with content inspector shown

We can now add buttons to the **Content** container. For now, we'll create just two buttons, for the two heads. Later, we'll expand it with more options. Each button will display an image icon (if you don't have an icon for your own content, you can use text labels):

1. In the **Hierarchy**, *right-click* the **Content** game object and select **UI | Button**. Rename it `PlasticHead Button`.

2. Set its **Width, Height** to (`150, 150`).

3. Remove its child **Text** object (unless you don't have an icon image for this button).

4. From the **Project** window, drag the `plastichead icon` image asset (perhaps found in your `/icons` folder) onto the **Image | Source Image** slot.

5. In the **Inspector**, click the + button at the bottom right of the **On Click** area of the **Button** component.

6. From the **Hierarchy**, drag the **Main Mode** object (under **Interaction Controller**), into the **Inspector**, and drop it onto the **On Click Object** slot.

7. In the **Function** selection list, choose **FaceMainMode | ChangePosePrefab**.

8. From the **Project** window, drag the **MrPlasticHead** prefab (in your `Prefabs/` folder) onto the empty parameter slot, as shown in the following screenshot:

Figure 9.5 – The PlasticHead button's On Click action will pass the MrPlasticHead prefab to the FaceMainMode.ChangePosePrefab function

> **Tip: Creating button icons**
>
> To create many of the button icons used in this chapter, I sometimes start by making a screen capture of the actual game object. Then, in Photoshop, I isolate the shape by selecting its edges (using the Magic Wand tool) and make a cutout with a transparent background. I then crop the image on a square-shaped canvas and resize it to 256x256, before exporting it as a PNG file. Then, in Unity, I import the image and, in **Import Settings**, set **Texture Type** to **Sprite (2D or UI)**, and click **Apply**. The asset can now be used as a UI sprite in an image component like those on button objects.

We now have one button in the **Main Menu**. This is for selecting the MrPlasticHead model. Let's make a second button, for the MrFacetHead prefab. To do that, we can duplicate and modify the first button, as follows:

1. In the **Hierarchy**, select the **PlasticHead Button** game object.

2. From the main menu, select **Edit | Duplicate** (or press *Ctrl + D*). Rename the copy to `FacetHead Button`.

3. From the **Project** window, drag the `facethead icon` asset onto the **Image |**
   Source Image slot.

4. From the **Project** window, drag the **MrFacetHead** prefab (in your `Prefabs/`
   folder) onto the parameter slot (replacing the **MrPlasticHead** prefab already there).

The **Main Menu** now has two buttons. When the app runs, clicking one will show
MrPlasticHead on my face. Clicking the other will show **MrFacetHead**. It would also be
nice to offer a reset button that clears all the face filters.

Adding a reset face button

We can also add a reset button that sets the current pose object to null. Let's do this as
a separate function in the `FaceMainMode` script. Use the following steps:

1. Open the `FaceMainMode` script for editing, and add a `ResetFace` function:

```
public void ResetFace()
{
    foreach (ARFace face in faceManager.trackables)
    {
        ChangeableFace changeable =
            face.GetComponent<ChangeableFace>();
        if (changeable != null)
        {
            changeable.SetPosePrefab(null);
        }
    }
}
```

2. In Unity, under **Content** in the main menu, *right-click* and select **UI | Button -**
   TextMeshPro. Rename it as `Reset Button`.

3. Set its **Width, Height** to (`150, 150`). Remove its **Image** component.

4. On its child **Text (TMP)** object, change the text to say `Reset`, check the **Auto Size**
   checkbox, and change the text **Vertex Color**, if you want.

5. Click the + button on the **On Click** list, drag the **Main Mode** object onto the **Object**
   slot, and select **FaceMainMode | ResetFace** from the **Function** list.

My main menu, at the bottom of the screen, now looks like this with its three buttons:

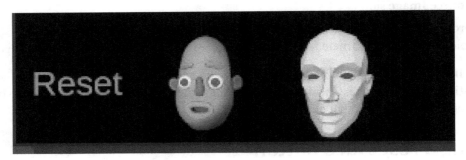

Figure 9.6 – The Main Menu with three buttons

You're now ready to build and run the project. Save your work (**File | Save**) and build it (**File | Build and Run**). You now have a little Face Maker app that lets you choose between 3D heads or **resetting the scene**!

In this section, we created a **Changeable Face Prefab** that you can set the child prefab of at runtime so the user can select different head models for their selfie. We then created a **Main Menu** panel with horizontally scrollable buttons, and added buttons that allow the user to choose **MrPlasticHead**, **MrFacetHead**, or to reset the current model.

Next, let's add some 3D accessories to your face – sunglasses and a hat.

Attaching 3D Accessories

Suppose you now want to accessorize your face and head. The setup is very similar to the pose prefabs we just used. For this, we will introduce a couple of third-party models downloaded from the web (and imported into your project at the top of this chapter). We'll also add an `AddAccessory` function to the **Changeable Face Prefab** that allows the user to view more than one accessory at a time.

Wearing a hat

I found a 3D hat on the internet (`https://free3d.com/3d-model/cartola-278168.html`), and we downloaded and installed it earlier in this chapter. Feel free to use this model and/or find your own model to add to the project. I installed it in my `Assets/Models/TopHat/` folder. The model is an FBX file named `CapCartola`. We'll also need to configure its materials.

If you select the `CapCartola` model in the **Project** window and unfold it, you'll notice it has child **Camera** and **Light** objects. This is not unusual for models exported from some 3D modeling programs such as Blender, for example. We obviously do not need these in our scene, so we'll also remove them from the imported model. Then we'll extract and set up the materials, and then put them together as a prefab. Follow these steps:

1. In the **Project** window, select the `CapCartola` model (in the `Assets/Models/TopHat/` folder).

2. In the **Inspector**, you'll see **Import Settings**. Make sure the **Model** tab is selected at the top of the window.

3. Uncheck the **Import Cameras** and **Import Lights** checkboxes. Then click **Apply**.

4. Select the **Materials** tab at the top of the **Inspector** window.

5. Click the **Extract Materials** button. This creates two new files, `Material.001` (for the hat itself) and `Material.002` (for its ribbon band). These are already associated with the model.

6. In the **Project** window, *right-click* your `Prefabs/` folder and select **Create | Prefab**. Rename it `TopHat`. Then open the prefab for editing.

7. From the **Project** window, drag the **CapCartola** model into the **Hierarchy**, creating a child instance under the root **TopHat** object.

8. With **CapCartola** selected in the **Hierarchy**, I found these **Transform** settings work for me: **Position**: `(0, 0.18, -0.02)`, **Rotation**: `(-20, 0, 0)`, and **Scale**: `(0.077, 0.077, 0.077)`.

9. Unfold **CapCartola** in the **Hierarchy** and select its child **Cylinder** object.

10. In the **Inspector**, under **Material.001** (for the hat itself), set the **Base Map** color to a blackish color (I used `#331D1D`).

11. Likewise, under **Material.002** (for the ribbon band), set the **Base Map** color to a nice red (I used `#FF1919`).

12. If you add an **AR Face** component to the root object, you can test it out right away by using it as the **AR Face Manager | Face Prefab**.

13. **Save** the prefab and exit back to the scene hierarchy.

You now have a **TopHat** prefab that you can use to accessorize your face. Let's also add sunglasses.

Sporting cool sunglasses

I found a 3D sunglasses model on the internet (`https://free3d.com/3d-model/sunglasses-v1--803862.html`), which we downloaded and installed earlier in this chapter. I installed it in my `Assets/Models/Sunglasses/` folder. The original model is an OBJ file named `12983_Sunglasses_v2_13`. We'll also need to configure its materials.

Extract and set up the materials, and then put the model together as a prefab using the following steps:

1. In the **Project** window, navigate to the folder containing the `12983_Sunglasses_v2_13` model and select it.

2. In the **Inspector**, you'll see **Import Settings**. Select the **Materials** tab at the top of the **Inspector** window and click the **Extract Materials** button. This creates two new files, `sunglasses_body` and `sunglasses_lens`.

3. Select the **sunglasses_body** material and adjust it as you desire. I made mine black. The lens material may be fine as is (dark with transparency).

4. In the **Project** window, *right-click* your `Prefabs/` folder and select **Create | Prefab**. Rename it `Sunglasses`.

5. Open the **Sunglasses** prefab for editing.

6. From the **Project** window, drag the **12983_Sunglasses_v2_l3** model into the **Hierarchy**, creating a child instance under the root **Sunglasses**.

7. With **12983_Sunglasses_v2_l3** selected in the **Hierarchy**, I found these **Transform** settings work for me: **Position**: (`-0.08`, `-0.025`, `-0.058`), **Rotation**: (`-90`, `90`, `09`), and **Scale**: (`0.0235`, `0.0235`, `0.0235`).

8. If you also add an **AR Face** component to the root object, you can test it out right away by using it as the **AR Face Manager | Face Prefab**.

9. Save the prefab and exit back to the scene hierarchy.

We now have two models we can use as face accessories. You can test them out by manually adding one to the **AR Session Origin | AR Face Manager | Face Prefab** slot and building and running the project. When you're done, don't forget to put the **Changeable Face Prefab** back into the slot.

Next, we'll add support for these accessories in the scripts.

Updating the scripts for accessories

We need to update the `ChangeableFace` script to manage the accessory objects. It will maintain a list of the current accessory objects, ensuring we create only one instance of any prefab.

Instead of destroying an accessory object when it's removed from the scene, we'll disable it, and then re-enable it if the user adds the same object again.

We also need to update the `FaceMainMode` script with a function that the menu buttons can call. This in turn passes the requested prefab to `ChangeableFace`.

Use the following steps to update your scripts:

1. Begin by opening the `ChangeableFace` script for editing and add the following declaration at the top of the class:

    ```
    Dictionary<GameObject, GameObject> accessories =
        new Dictionary<GameObject, GameObject>();
    ```

    We're using a dictionary to maintain the list of instantiated accessory objects, keyed by the prefab.

2. Then, add an `AddAccessory` function as follows:

    ```
    public void AddAccessory(GameObject prefab)
    {
        GameObject obj;
        if (accessories.TryGetValue(prefab, out obj) &&
            obj.activeInHierarchy)
        {
            obj.SetActive(false);
            return;
        }
        else if (obj != null)
        {
            obj.SetActive(true);
        }
        else
        {
            obj = Instantiate(prefab, transform, false);
    ```

```
            accessories.Add(prefab, obj);
    }
}
```

AddAccessory instantiates the prefab as a child of the face and adds it to the accessories list. However, if the prefab has already been instantiated, we remove it from the scene by setting it as inactive. Likewise, if you try to add it again, it is reactivated.

3. Next, we'll add a ResetAccessories function that removes all accessories, as follows:

```
public void ResetAccessories()
{
    foreach (GameObject prefab in accessories.Keys)
    {
        accessories[prefab].SetActive(false);
    }
}
```

> **Tip: Avoid garbage collection by using object caching**
>
> In this AddAccessory function, I could have called Destroy to remove an existing instance, and then called Instantiate again if and when the object was added a second time. Instead, I'm managing memory by simply disabling existing objects when not wanted and reusing the same instances when requested. Repeatedly instantiating and destroying objects at runtime can cause memory fragmentation and require Unity to perform memory **garbage collection (GC)**. GC can be computationally expensive and cause glitches in your frame rate updates (see https://docs.unity3d.com/Manual/UnderstandingAutomaticMemoryManagement.html). Likewise, you may want to refactor the other scripts in this chapter to not use Destroy.

4. Next, we can open the FaceMainMenu script for editing, and add an AddAccessory function that will be called by the menu buttons, as follows:

```
public void AddAccessory(GameObject prefab)
{
    foreach (ARFace face in faceManager.trackables)
    {
        ChangeableFace changeable =
            face.GetComponent<ChangeableFace>();
```

```
              if (changeable != null)
              {
                   changeable.AddAccessory(prefab);
              }
          }
      }
```

5. Next, add the following highlighted code to ResetFace:

```
      public void ResetFace()
      {
          foreach (ARFace face in faceManager.trackables)
          {
              ChangeableFace changeable =
                  face.GetComponent<ChangeableFace>();
              if (changeable != null)
              {
                  changeable.SetPosePrefab(null);
                  changeable.ResetAccessories();
              }
          }
      }
```

We're now ready to add menu buttons for the TopHat and Sunglasses accessories.

Adding accessories to the main menu

To add new buttons to the main menu, we can duplicate an existing button and modify it by following these steps:

1. In the **Hierarchy**, duplicate one of the menu buttons (found under **Main UI | MainMenu Panel | Scroll View | Viewport | Content**), such as the **FacetHead Button** game object. On the main menu, click **Edit | Duplicate** (or press *Ctrl + D*). Rename the copy to HatAccessory Button.

2. From the **Project** window, drag the tophat icon asset onto the **Image | Source Image** slot.

3. On the **Button On Click** action, change **Function** to FaceMainMode. AddAccessory.

4. From the **Project** window, drag the **TopHat** prefab (in your `Prefabs/` folder) onto the parameter slot.

5. Likewise, repeat *steps 1-4* for a `SunglassesAcessory Button`, using the `sunglasses icon` image and the **Sunglasses** prefab asset.

Save the scene and build the project. When you tap the hat button, you're wearing a top hat. Tap it again to remove it. In the following screen capture, I'm wearing the facet face, top hat, and sunglasses. I've never looked so cool!

Figure 9.7 – Selfie with me wearing a top hat, sunglasses, and faceted face at the same time

In this section, we built upon the basic face pose tracking features by adding other models to be tracked at the same time. We created prefabs for the TopHat and Sunglasses using models download from the web. Then, we updated the `ChangeableFace` script to handle multiple accessory objects. This implements good memory management practices by avoiding duplicate instances of the same prefab and caching the spawned instances in a dictionary list. After updating the `FaceMainMode` script with a public `AddAccessory` function, we added new buttons to the main menu so that the user can accessorize their head with a hat and/or sunglasses.

So far, all our faces are fixed-expression static models. AR Foundation also supports the dynamic visualization of faces. Let's try that next.

Making dynamic face meshes with a variety of materials

To show an augmented face that matches your real-life expressions, Unity AR Foundation lets you generate a face mesh dynamically at runtime. On this mesh, you can apply different materials, giving the effect of you wearing arbitrary face masks. To add this to our project, we'll first look at the default face game object given by AR Foundation. Then we'll create several different materials to use. To integrate this feature into our project, we'll extend the `ChangeableFace` script to switch materials, add a similar function to the `FaceMainMode` script to update the face trackables, and then add menu buttons to switch materials.

Exploring AR Default Face

You can create a dynamic face game object for AR Foundation from the Unity menu at **GameObject | XR | AR Default Face**. The object includes an **AR Face Mesh Visualizer** component that generates a face mesh at runtime that matches your facial expressions, including moving your mouth and raising your eyebrows. Let's quickly try it out before we add this feature to our **Changeable Face Prefab**. Use the following steps:

1. From the Editor menu bar, select **GameObject | XR | AR Default Face**. This creates an object named **AR Default Face** in the scene hierarchy.

   Note that you won't see this object in your Scene window because the mesh is dynamically generated at runtime, so there's nothing to render yet.

2. Replace the default material (the one included is not for URP): In the **Project** window, *right-click* your `Materials/` folder (create one first if necessary), and name it `DefaultFace Material`. Set the **Base Map** color to your favorite color. Drag the material onto the **AR Default Face** object.

3. Make it a prefab. Drag the **AR Default Face** object into the **Project** window's `Prefabs/` folder.

4. Then delete it from the **Hierarchy**.

5. Now, drag the prefab onto your **AR Session Origin | AR Face Manager | Face Prefab** slot.

Here's a screen capture of me wearing the default mask, and smiling brightly, on the left. On the right is a scene view of my face mesh generated at runtime:

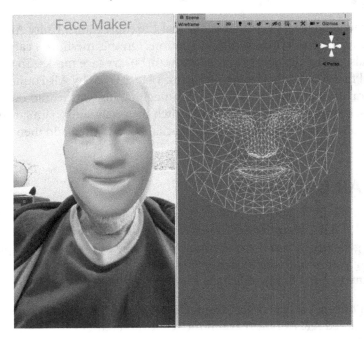

Figure 9.8 – Me wearing an AR default mask (left) and a wireframe of my face mesh (right)

It's easy to replace this default material with other materials to make your own masks.

Creating face materials

For fun (and for the purposes of instruction), let's try using an arbitrary photo as a face texture. I'll use a picture named `WinterBarn.jpg` (this was also used in *Chapter 6*, *Gallery: Building an AR App*). Create a new material using the photo, with the following steps:

1. *Right-click* in your **Project** window `Materials/` folder and select **Create | Material**, naming it `PhotoFace Material`.

2. Drag a photo from the **Project** window (for example, `WinterBar.jpg`) onto the **Base Map** texture chip. Ensure the **Base Map** color is white.

3. Duplicate the **AR Default Face** prefab by selecting it in the **Project** window and choosing **Edit | Duplicate** (or pressing *Ctrl + D*). Then rename it `PhotoFace Prefab`.

4. Open the new prefab for editing and drag the **PhotoFace Material** onto it. **Save** the prefab and return to the scene hierarchy.

5. To try it out, drag the **PhotoFace Prefab** onto **AR Face Manager | Face Prefab** and run the scene.

This ought to give you a feeling of how a 2D texture image is mapped onto the face mesh. This is called **UV mapping**. In the following figure, I'm wearing a mask with this ordinary photo as a texture map. On the right is an actual UV texture map (`PopFace_Albedo`) for the face mesh:

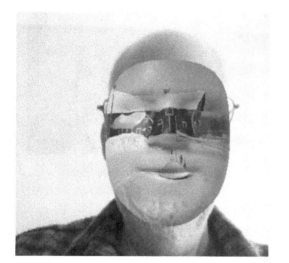 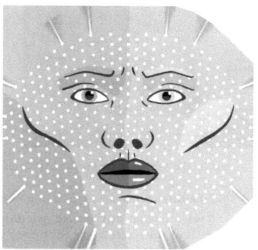

Figure 9.9 – Ordinary 2D image as face texture (left), and a UV mapped face texture (right)

In this way, you can use any 2D photograph or image that you want. Try others for yourself, such as your national flag, the logo of your favorite sports team, and so on.

The `PopFace_Albedo` texture shown in the preceding figure is included in the AR Face Assets package from Unity that we imported into our project at the beginning of this chapter. Make a material for that now by repeating *steps 1-5*, naming the material `PopFace Material`, and using `PopFace_Albedo` for the **Base Map** texture.

Likewise, the AR Face Assets package includes textures for a robot face. Again, repeat *steps 1-5* for a new `RobotFace Material`, using `Robot_Albedo` for the **Base Map** texture. In this case, there are additional texture maps you should also add to the material – `Robot_Normal` and `Robot_Occlusion` for **Normal Map** and **Occlusion Map**, respectively.

When adding the **Normal Map** texture, you may be prompted with **This texture is not marked as a normal map**. Click the **Fix Now** button to apply the required **Import Settings**.

The following figure shows me wearing the **RobotFace** and **PopFace** masks. Not obvious in these screen captures is the fact that the face mesh follows my facial expressions in real time:

Figure 9.10 – Selfies using the Robot PBR material (left), and the Pop albedo texture (right)

> **Info: Using Procreate to paint your own textures**
>
> If you're interested in painting your own UV mapped face textures (and have an iPad), the Procreate app (https://procreate.art/) has a feature for doing this (check out *Dilmer Valecillos's* video on this at https://youtu.be/FOxhcRzDLx8).

With the materials made, we can add the face mesh visualizer to the changeable face prefab, so it will generate the face mesh at runtime.

Adding a face mesh visualizer to the changeable face prefab

To integrate a dynamic face mesh into our app, we should add it to our versatile **Changeable Face Prefab**. We will need the same components as the **AR Default Face** game object we generated earlier, and they need to be on the prefab's root object. Use the following steps to add them manually:

1.   Open **Changeable Face Prefab** for editing.
2.   With the prefab root object selected, click **Add Component** in the **Inspector**.

3. Search for and select the **AR Face Mesh Visualizer** component.

4. Search for and select a **Mesh Filter** component.

5. Search for and select a **Mesh Renderer** component.

6. Drag the **DefaultFace Material** from the **Project** window onto the **Changeable Face Prefab** root object.

7. Save the prefab.

8. Back in the scene hierarchy, drag the **Changeable Face Prefab** asset onto the **AR Session Origin | AR Face Manager | Face Prefab** slot.

If you build and run now, you'll see the default face mesh. All the menu buttons still work, letting you add 3D head models and accessories.

We want to have buttons that let the user choose between face materials. For that, we need to update our scripts.

Controlling the face material

We can hide or show the face mesh by toggling the **AR Face Mesh Visualizer** and **Mesh Render** components. Use the following steps:

1. Open the `ChangeableFace` script for editing and add the following at the top of the script:

    ```
    using UnityEngine.XR.ARFoundation;
    ```

2. Add the following code to declare and initialize references to the `ARFaceMeshVisualizer` and `MeshRenderer` components:

    ```
    ARFaceMeshVisualizer meshVisualizer;
    MeshRenderer renderer;

    private void Start()
    {
        meshVisualizer =
            GetComponent<ARFaceMeshVisualizer>();
        meshVisualizer.enabled = false;
        renderer = GetComponent<MeshRenderer>();
        renderer.enabled = false;
    }
    ```

    We'll start the app with the face mesh not visible, so both components are disabled.

3. Then, add a `SetMeshMaterial` function as follows:

```
public void SetMeshMaterial(Material mat)
{
    if (mat == null)
    {
        meshVisualizer.enabled = false;
        renderer.enabled = false;
        return;
    }
    renderer.material = mat;
    meshVisualizer.enabled = true;
    renderer.enabled = true;
}
```

When given a material, `mat`, the function sets it in the renderer and makes sure the visualizer and renderer components are enabled. If you pass a `null` value for the `mat`, then the components will be disabled.

4. Next, open the `FaceMainMode` script and add a `ChangeMaterial` function, as follows:

```
public void ChangeMaterial(Material mat)
{
    foreach (ARFace face in faceManager.trackables)
    {
        ChangeableFace changeable =
            face.GetComponent<ChangeableFace>();
        if (changeable != null)
        {
            changeable.SetMeshMaterial(mat);
        }
    }
}
```

Like the other functions in the script, it loops through any trackables and calls into the changeable component.

5. Next, update the `ResetFace` function with the following highlighted line:

```
changeable.SetPosePrefab(null);
changeable.ResetAccessories();
changeable.SetMeshMaterial(null);
```

The code is now written. We added a `SetMaterial` function to the `ChangeableFace` script that enables the mesh visualizer and sets the material to render. To the `FaceMainMode` script, we added a `ChangeMaterial` function that calls `SetMaterial` on each trackable AR face.

We're now ready to add menu buttons for the various mesh materials.

Adding face materials to the main menu

To add new buttons to the main menu, we can duplicate an existing button and modify it, as we did earlier. Use the following steps:

1. In the **Hierarchy**, duplicate one of the menu buttons (found under **Main UI** | **MainMenu Panel** | **Scroll View** | **Viewport** | **Content**), such as the **FacetHead Button** game object, using the main menu **Edit** | **Duplicate options** (or press *Ctrl + D*). Rename the copy to `DefaultFace Button`.

2. From the **Project** window, drag the `default face icon` asset onto the **Image** | **Source Image** slot.

3. On the button **On Click** action, change **Function** to **FaceMainMode. ChangeMaterial**.

4. From the **Project** window, drag the **DefaultFace Material** (in your `Materials/` folder) onto the parameter slot.

5. Likewise, repeat *steps 1-4* three times, for `PhotoFace Button` (using the `photo face icon` image, and the **PhotoFace Material** asset), for `PopFace Button`, and for `RobotFace Button`.

Save the scene and build the project. When you tap one of the face material buttons, it renders the face mesh. The following cropped screen capture shows the horizontally-scrolled menu with the new buttons:

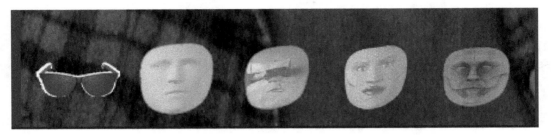

Figure 9.11 – Face mesh texture buttons on the main menu

In this section, we added an **AR Face Mesh Visualizer** component to our **Changeable Face Prefab** so that face meshes will be generated at runtime. We created several materials to apply when rendering the mesh, and then added buttons to the main menu that let the user choose between materials. When a button is clicked, it sends the material asset to the FaceMainMode. This then forwards it to the trackable face(s).

While the face visualizer can follow some of your expressions, including raised eyebrows and opening your mouth, it does nothing for your eyes. Let's consider eye tracking next.

Using eye tracking (ARKit)

For eye tracking, as you might expect, you are given the pose transforms for each eye, which you can use to update your own "eyeball" game objects. For this feature, I'll show you how to do it, but leave the details of integrating it into the project up to you. Presently, this feature requires an iOS device with a **TrueDepth** camera.

To learn more about eye tracking with AR Foundation, take a look at the EyeLasers scene given in the AR Foundation sample assets (we installed these in the Assets/ARF-samples/ folder).

The **Face Prefab** in the scene's **AR Face Manager** is the **AR Eye Laser Visualizer** prefab. This has an **AR Face** component (as you would expect), plus an **Eye Pose Visualizer**. This visualizer script, in turn, is given an eyeball prefab. In this specific scene, it is given the **Eye Laser Prefab**. This simply contains a long thin cylinder that'll be rendered to look like a laser beam. In summary, these dependencies could be depicted as the following:

EyeLasers scene -> AR Eye Laser Visualizer face prefab -> Eye Pose Visualizer script -> Eye Laser Prefab

The `EyePoseVisualizer` script is an example script (not part of the AR Foundation package itself). Briefly, you give it an eyeball prefab, which is instantiated twice and parented by the `ARFace`, `leftEye`, and `rightEye` pose transforms. For example, you'll find this line of code in the script's `CreateEyeGameObjectsIfNecessary` function (line 45):

```
m_LeftEyeGameObject = Instantiate(m_EyePrefab, m_Face.leftEye);
```

As a child of the tracked eye transforms, the spawned objects appear to automatically track with your detected eye movements.

The script also subscribes to the `ARFace` and `update` events, where it toggles the eyes' visibility based on the trackable's tracking state, as shown in the following code:

```
void OnUpdated(ARFaceUpdatedEventArgs eventArgs)
{
    CreateEyeGameObjectsIfNecessary();
    SetVisible((m_Face.trackingState ==
        TrackingState.Tracking) &&
            (ARSession.state > ARSessionState.Ready));
}
```

> **Tip: Using updated events with face tracking**
>
> This script illustrates another best practice for face tracking with AR Foundation. By subscribing to the trackables' `updated` events, it toggles the visibility of instantiated prefabs based on the trackable's `trackingState`, as well as the overall `ARSession.state`. You might consider refactoring the functions in our `FaceMainMode` class to handle `updated` events in this way too.

Eye tracking is not available on all platforms. When the script is enabled, it first checks the Unity eye tracking subsystem. If the feature is not supported, the component disables itself, as highlighted in the following `OnEnable` function (lines 65-78):

```
void OnEnable()
{
    var faceManager =
        FindObjectOfType<ARFaceManager>();
    if (faceManager != null && faceManager.subsystem !=
    null && faceManager.descriptor.supportsEyeTracking)
    {
```

```
        m_FaceSubsystem =
            (XRFaceSubsystem)faceManager.subsystem;
        SetVisible((m_Face.trackingState ==
            TrackingState.Tracking) &&
            (ARSession.state > ARSessionState.Ready));
        m_Face.updated += OnUpdated;
    }
    else
    {
        enabled = false;
    }
}
```

If you want to try this yourself with an eyeball instead of a laser beam, the following URL contains a free eyeball 3D model you could use: `https://free3d.com/3d-model/eyeball--33237.html`. Make it into a prefab and substitute it for the eye laser prefab on the AR eye laser visualizer prefab's **Eye Pose Visualizer | Eye Prefab** slot.

This is fantastic! However, you can do so much more. For example, with ARCore, you can attach graphics to specific regions of the face. Let's look into that now.

Attaching stickers to face regions (ARCore)

If your project is using ARCore XR Plugin and Android, you have access to ARCore-specific features, including transforms for three important face regions: the nose tip, left forehead, and right forehead. If you raise your left eyebrow, for example, that transform will move independently of the rest of the face, providing some more detail to the facial expressions in your app.

In addition to what we do here, you may also want to look at the **ARCoreFaceRegions** scene in the AR Foundation Samples project (see the `ARF-samples/` folder in your project), and the `ARCoreFaceRegionManager` script it uses. The code we develop in this section is considerably simpler and easier to follow.

To demonstrate ARCore face regions, we'll implement several 2D stickers and attach them to the 3D face regions. We'll let you add eyebrows, a mustache, and licking lips using clipart that we identified at the top of this chapter (and I edited in Photoshop). They've been imported as **Sprite (2D and UI)**. These are available in this book's GitHub repository.

We can start by creating the sticker prefabs.

Creating the sticker prefabs

To make prefabs of these clipart images, use the following steps:

1. *Right-click* in the **Project** window and select **Create | Prefab**. Rename it `Mustache Prefab`. Then open it for editing.

2. From the **Project** window, drag the `mustache` image onto the root **Mustache Prefab**. This creates a child object named **mustache** with a **Sprite Renderer** component.

3. Set the **mustache** object's **Transform**. The following values worked for me: **Position**: `(0, -0.02, 0)` and **Scale**: `(0,019, 0,019, 0,019)`.

4. Save the prefab.

5. Repeat *steps 1-4*, making `Lips Prefab` using the `licking-lips` sprite image. Use **Position**: `(0, -0.05, 0)` and **Scale**: `(0,019, 0,019, 0,019)`.

6. Again, repeat *steps 1-4*, making `Eyebrow Left Prefab` using the `eyebrow-left` sprite image. Use **Position**: `(0, -0.01, 0)` and **Scale**: `(0,019, 0,019, 0,019)`.

7. And likewise, one more time, make `Eyebrow Right Prefab` using the `eyebrow-right` sprite image. Use **Position**: `(0, -0.01, 0)` and **Scale**: `(0,019, 0,019, 0,019)`.

We now have prefabs for a mustache, lips, and eyebrows. Let's write the scripts to attach them using the ARCore face regions support.

Managing attachments' positions

We'll create a separate script, `FaceRegionAttachments` on **Changeable Face Prefab**. It makes sense to keep it separate from the `ChangeableFace` script because the code is ARCore-specific and is relatively long.

The lines of code that depend on ARCore are enclosed in `#if UNITY_ANDROID && !UNITY_EDITOR` compiler symbols, so they will not run in a non-Android environment (including the desktop play mode). Use the following steps:

1. Create a new C# script named `FaceRegionAttachments` and open it for editing.

2. Start writing the script by replacing the content with the following code:

```
using System.Collections.Generic;
using UnityEngine;
using Unity.Collections;
using UnityEngine.XR.ARFoundation;
```

```
#if UNITY_ANDROID
using UnityEngine.XR.ARCore;
#endif

public class FaceRegionAttachments : MonoBehaviour
{
    ARFaceManager faceManager;
    ARFace face;

    Dictionary<ARCoreFaceRegion, GameObject> prefabs =
        new Dictionary<ARCoreFaceRegion, GameObject>();
    Dictionary<ARCoreFaceRegion, GameObject> objs =
        new Dictionary<ARCoreFaceRegion, GameObject>();

#if UNITY_ANDROID && !UNITY_EDITOR
    NativeArray<ARCoreFaceRegionData> faceRegions;
#endif

    private void Start()
    {
        faceManager = FindObjectOfType<ARFaceManager>();
        face = GetComponent<ARFace>();
    }
```

The script first declares that we're using the ARFoundation API as well as ARCore. Then, at the top of the class, we declare variables for ARFaceManager and the object's ARFace, and initialize these in Start. We also declare two dictionaries, prefabs and objs, that will be indexed by the ARCore region identifier (enum). We then declare a NativeArray of ARCoreFaceRegionData named faceRegions that we'll be using in Update.

3. Add a SetRegionAttachment function (that will be called from FaceMainMode) as follows:

```
    public void SetRegionAttachment(ARCoreFaceRegion
        region, GameObject prefab)
    {
        GameObject obj;
        if (objs.TryGetValue(region, out obj))
        {
```

```
        GameObject currentPrefab = prefabs[region];
        Destroy(obj);
        prefabs.Remove(region);
        objs.Remove(region);
        if (prefab == currentPrefab)
            return;
    }

    obj = Instantiate(prefab);
    prefabs.Add(region, prefab);
    objs.Add(region, obj);
}
```

The function gets a `region` ID and a `prefab`, instantiates the `prefab`, and records both the `prefab` and spawned object in the dictionaries. If there is already a spawned object, it is first destroyed and removed from the lists. We check if the new prefab was the same as the current one, so it won't be respawned again, effectively allowing the menu button to toggle on and off as an attachment by clicking twice.

4. On each `Update`, we need to ask ARCore for the current list of face regions, and update the spawned object transforms accordingly, as follows:

```
    private void Update()
    {
#if UNITY_ANDROID && !UNITY_EDITOR
        var subsystem =
            (ARCoreFaceSubsystem)faceManager.subsystem;
        if (subsystem == null)
            return;

        subsystem.GetRegionPoses(face.trackableId,
            Allocator.Persistent, ref faceRegions);
        for (int i = 0; i < faceRegions.Length; ++i)
        {
            GameObject obj;
            if (objs.TryGetValue(faceRegions[i].region,
                out obj))
            {
                obj.transform.localPosition =
                    faceRegions[i].pose.position;
```

```
        }
    }
#endif
    }
```

5. We can also provide a public `Reset` function that destroys all the instantiated objects and clears the dictionaries:

```
public void Reset()
{
    foreach (ARCoreFaceRegion region in objs.Keys)
    {
        Destroy(objs[region]);
    }
    objs.Clear();
    prefabs.Clear();
}
```

6. Finally, it's good practice to dispose of the `faceRegions` native array when this game object is destroyed, as follows:

```
void OnDestroy()
{
#if UNITY_ANDROID && !UNITY_EDITOR
    if (faceRegions.IsCreated)
        faceRegions.Dispose();
#endif
    }
}
```

7. Save the script, then, back in Unity, open the **Changeable Face Prefab** asset for editing.

8. Drag the `FaceRegionAttachments` script onto the root **Changeable Face Prefab** game object, then save the prefab.

Tip: Refactor to avoid garbage collection

As we did earlier, in the *Attaching 3D Accessories* section, you may want to refactor this code to avoid garbage collection by reusing objects rather than repeatedly calling `Destroy` and `Instantiate` for the same prefabs.

Now we'll update the `FaceMainMode` script to use it and provide public functions that the menu buttons can call, as follows:

1. Open the `FaceMainMode` script for editing, and start by adding the following lines at the top of the file (needed for the enum `ARCoreFaceRegion` definition):

    ```
    #if UNITY_ANDROID
    using UnityEngine.XR.ARCore;
    #endif
    ```

2. Add a private `SetRegionAttachment` function that loops through the trackables and calls `SetRegionAttachment` on them:

    ```
    private void SetRegionAttachment(ARCoreFaceRegion
        region, GameObject prefab)
    {
        foreach (ARFace face in faceManager.trackables)
        {
            FaceRegionAttachments regionAttachments =
              face.GetComponent<FaceRegionAttachments>();
            if (regionAttachments != null)
            {
                regionAttachments.
                    SetRegionAttachment(region, prefab);
            }
        }
    }
    ```

3. Next, expose this capability via separate public functions we can call from the menu button Unity actions, as follows:

    ```
    public void SetNoseAttachment(GameObject prefab)
    {
        SetRegionAttachment(ARCoreFaceRegion.NoseTip,
            prefab);
    }


    public void SetForeheadLeftAttachment(GameObject
        prefab)
    {
    ```

```
        SetRegionAttachment(
            ARCoreFaceRegion.ForeheadLeft, prefab);
    }

    public void SetForeheadRightAttachment(GameObject
        prefab)
    {
        SetRegionAttachment(
            ARCoreFaceRegion.ForeheadRight, prefab);
    }
```

4. Save the script.

Here, we created a new `FaceRegionAttachments` script that maintains dictionary lists of `prefabs` and spawned `objs` for game objects attached to specific face regions. On each frame `Update`, the `objs` transforms are updated based on the face region's pose transform, so it tracks with its region. This implementation allows multiple attachments on a face, but only one per region. Then, we updated the `FaceMainMode` script with public functions that can be called by menu buttons to add attachments.

We can now make the menu buttons.

Adding region attachments to the main menu

As we did earlier, to add new buttons to the main menu, we can duplicate an existing button and modify it. Use the following steps:

1. In the **Hierarchy**, duplicate one of the menu buttons (found under **Main UI | MainMenu Panel | Scroll View | Viewport | Content**), such as the **FacetHead Button** game object. Using the main menu, navigate to **Edit | Duplicate** (or press *Ctrl + D*). Rename the copy to `Mustache Button`.

2. From the **Project** window, drag the `mustache icon` asset onto the **Image | Source Image** slot.

3. On the button **On Click** action, change **Function** to **FaceMainMode. SetNoseAttachment**.

4. From the **Project** window, drag the **Mustache Prefab** asset onto the parameter slot.

5. Repeat *steps 1-4* for `Lips Button`, using the `licking-lips icon` image, and the **Lips Prefab** asset. Use the same function as the mustache, **FaceMainMode. SetNoseAttachment**.

6. Repeat *steps 1-4* again for `Eyebrows Button`, using the `eyebrows icon` image. This time, we'll have two **On Click** actions, one for each eye. The first calls **FaceMainMode.SetForeheadLeftAttachment** with the **EyebrowLeft Prefab**. The second calls **FaceMainMode.SetForeheadRightAttachment** with the **EyebrowRight Prefab**, as shown in the following:

Figure 9.12 – The eyebrows button has two On Click actions, for both the left and right regions and prefabs

Save the scene and build the project. When you tap one of the region attachment buttons, it adds its sticker prefabs to the scene. The mustache and lips both set the nose attachment so you can only view one at a time. The following screen captures show me all decked out, including combining it with other face augmentations we created earlier (right):

Figure 9.13 – Selfie screenshots with multiple stickers, and (on the right) combined with other augmentations

Because this feature is specific to ARCore, you will probably want to hide the sticker buttons if you try building the project for iOS. We can add those next.

ARCore-only UI buttons

This face region stickers feature only runs on ARCore and Android. If you plan to build the same project on iOS (as well as Android), we already account for code compilation issues using conditional compile symbols. However, the menu buttons will still be visible. You could disable them by hand in the editor before doing a build, or you could let a script handle it.

Use the following `ARCoreOnly` script to hide buttons from the UI (unless you're targeting Android). If you're targeting Android but using play mode in the editor (using the AR Foundation remote tool), this script will disable the button so that it is visible but not interactable:

```
using UnityEngine;
using UnityEngine.UI;

public class ARCoreOnly : MonoBehaviour
{
    private void Awake()
    {
#if !UNITY_ANDROID
        gameObject.SetActive(false);
#endif
#if UNITY_EDITOR
        Button button = GetComponent<Button>();
        button.interactable = false;
#endif
    }
}
```

Drag a copy of this script onto the mustache button, lips button, and eyebrows button game objects so that they can only be used with ARCore.

To summarize, in this section, we created several sticker prefabs containing **Sprite Renderers**. We wrote an ARCore-specific script, `FaceRegionAttachments`, that uses the native `ARCoreFaceRegionData` (via `ARCoreFaceSubsystem`) to find the pose transform of each face region (nose tip, left forehead, and right forehead), and track each spawned game object with the given face region. We added menu buttons for each of the stickers that call public functions in `FaceMainMenu` by passing the sticker prefab to add. This in turn forwards the prefab to the trackable faces. Feel free to add more sticker prefabs and buttons, using similar steps to the ones found in this section.

This is cool, but having just three face regions is kind of limited. Using ARKit, you have access to much more refined detail about face geometry. This is achieved with the use of blend shapes.

Tracking expressive face blend shapes (ARKit)

ARKit introduces additional advanced face tracking features available only on iOS devices, including blend shapes. **Blend shapes** refer to morphing mesh geometries that are commonly used for animating the faces of NPCs (non-player characters) in video games and VR applications. Presently, they are an ARKit-specific feature. ARKit blend shapes provide intricate details of facial expressions as separate features, such as a left or right eye blink, looking down, eyes wide open, cheek puff, cheek squint, jaw left, mouth dimple, and many more. Each feature is given a coefficient on a scale of 0.0 to 1.0. This shape data can be forwarded to the Unity **Skinned Mesh Renderer** (`https://docs.unity3d.com/Manual/class-SkinnedMeshRenderer.html`) that is used in character animation. A good explanation and conversation can be found at the following URL: `https://www.quora.com/What-is-blendshape-exactly`.

Building an animated rig (with bones and a skinned mesh) is beyond the scope of this book. Instead, by way of explanation, I'll walk through the example assets given in the AR Foundation samples project's `ARKitFaceBlendShapes` scene, found in the `Assets/ARF-samples/scenes/FaceTracking/` folder. To begin, you can try it yourself (if you're set up for iOS development) by building the `ARKitFaceBlendShapes` scene. Now, let's take a closer look.

Opening the scene in the Unity Editor, you will find **AR Session Origin** has an **AR Face Manager** component, as you'd expect. This references the `SlothHead` prefab for the **Face Prefab**.

Opening the **SlothHead** prefab, you will see that its root game object has an **AR Face** component. It also has an `ARKitBlenShapeVisualizer`. This is an example script provided with AR Foundation samples (it is not part of the AR Foundation package itself). This component has a parameter for **Skinned Mesh Renderer**. This is on the **Sloth_Head2** child object, as shown in the following screenshot:

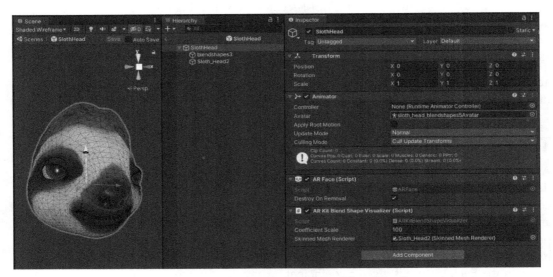

Figure 9.14 – The SlothHead prefab has the sample ARKitBlendShapeVisualizer script that references the skinned mesh render on the child Sloth_Head2

Open the `ARKitBlendShapeVisualizer` script in your code editor. You'll find a function, `CreateFeatureBlendMapping`, that is called `Awake`. This maps ARKit blend shape names (type `ARKitBlendShapeLocation`) with corresponding indexes on the `skinnedMeshRenderer`. For the list of locations and descriptions, see the following URL: `https://docs.unity3d.com/Packages/com.unity.xr.arkit-face-tracking@4.2/api/UnityEngine.XR.ARKit.ARKitBlendShapeLocation.html`.

The following screenshot shows the **Sloth_Head2** object's **Skinned Mesh Renderer**, with some of its **BlendShapes** visible in the Unity **Inspector**:

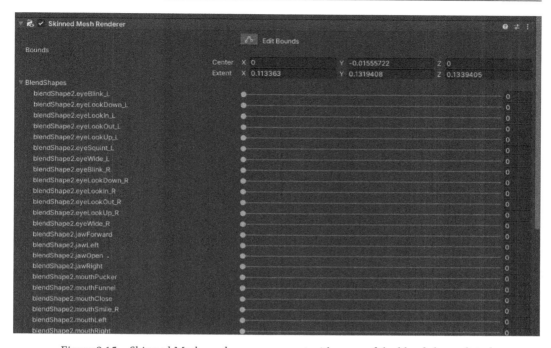

Figure 9.15 – Skinned Mesh renderer component with some of the blend shapes listed

The ARKit blend shape locations are mapped to the **Skinned Mesh Renderer** ones.

Then the `ARKitBlendShapeVisualizer` script, which subscribes to the `ARFace` `updated` events with the `OnUpdated` function, in turn, calls its `UpdateFaceFeatures` function. `UpdateFaceFeatures` gets the current blend shape coefficients from ARKit (`m_ARKitFaceSubsystem.GetBlendShapeCoefficients`), and for each coefficient, sets that coefficient value (scaled by a global scalar) to the `skinnedMeshRender`. From there, Unity does its magic, deforming and animating the mesh geometry to be rendered on the screen. This is not simple but does make sense if you can follow it correctly.

That's basically how blend shapes work. Developing your own models and code may require a good familiarity with the parts of Unity in question, but all the information you need is accessible. You will be successful if you know how to use it.

Summary

In this chapter, you built a face maker app that handles face tracking with the forward-facing (user-facing) camera on a mobile device. You learned that the **AR Face Manager** component takes a **Face Prefab** to instantiate when a face is tracked. You first used that to visualize specific **AR Face** prefabs but then created a generic **Changeable Face Prefab** with a `ChangeableFace` script that we could update from the `FaceMainMode` script.

You used this architecture to explore several ways of rendering tracked faces. First, you used the face pose to render an instantiated 3D head model (**MrPlasticHead** and **MrFacetHead**). Next, you used this technique to add accessories to the face, including a **TopHat** and **Sunglasses**. Then, you added an **AR Face Mesh Visualizer** to generate a face mesh dynamically at runtime, and then made several materials that can be applied to the mesh to make a wide variety of face masks. If you're on ARCore, you also implemented face region stickers using sprite images attached to ARCore face regions. Finally, you learned about ARKit-specific face tracking features, including eye tracking and blend shapes. In the process, you implemented a horizontally-scrolling main menu button that lets users choose various combinations of face filters. All this was great fun!

You now have a working knowledge of how to build AR applications in Unity using AR Foundation. If you followed along with each of the chapters of this book, you will have learned how to set up your system for AR development with Unity configured to build on your target platform and mobile device. You created a simple AR scene, learning the main game objects required for AR, including the AR Session and AR Session Origin. You also explored the sample AR projects provided by Unity. Next, you learned about improving the developer workflow and troubleshooting your apps, considering situations unique to AR development.

You created a user framework for developing an AR application that included onboarding graphics, interaction modes, and UI panels. This was saved as a scene template for reuse. You learned how to use the framework, first building a simple place-object scene with a simple main menu.

In the third part of the book, you built several AR applications, including a picture gallery that lets you place framed photos on your walls, with menus and user interactivity. You improved the app, adding editing tools to move, resize, delete, and change the images displayed in virtual pictures in the scene. In the next project, you used image tracking to present 3D graphics and information about the planets using real-life printed flashcards. Finally, in this chapter, you built a face tracking app with a scrolling menu containing a variety of face heads, masks, and attachable accessories to make fun selfies.

This is just the start. AR Foundation and Unity provide even more support for augmented reality applications, including object tracking and geotagging with GPS, as well as the full richness of the Unity platform for the development of interactive 3D games and applications. Go out and augment the world!

Packt.com

Subscribe to our online digital library for full access to over 7,000 books and videos, as well as industry leading tools to help you plan your personal development and advance your career. For more information, please visit our website.

Why subscribe?

- Spend less time learning and more time coding with practical eBooks and Videos from over 4,000 industry professionals

- Improve your learning with Skill Plans built especially for you

- Get a free eBook or video every month

- Fully searchable for easy access to vital information

- Copy and paste, print, and bookmark content

Did you know that Packt offers eBook versions of every book published, with PDF and ePub files available? You can upgrade to the eBook version at packt.com and as a print book customer, you are entitled to a discount on the eBook copy. Get in touch with us at customercare@packtpub.com for more details.

At www.packt.com, you can also read a collection of free technical articles, sign up for a range of free newsletters, and receive exclusive discounts and offers on Packt books and eBooks.

Other Books You May Enjoy

If you enjoyed this book, you may be interested in these other books by Packt:

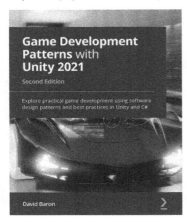

Game Development Patterns with Unity 2021 - Second Edition

David Baron

ISBN: 978-1-80020-081-4

- Structure professional Unity code using industry-standard development patterns
- Identify the right patterns for implementing specific game mechanics or features
- Develop configurable core game mechanics and ingredients that can be modified without writing a single line of code
- Review practical object-oriented programming (OOP) techniques and learn how they're used in the context of a Unity project
- Build unique game development systems such as a level editor
- Explore ways to adapt traditional design patterns for use with the Unity API

Unity 2020 Virtual Reality Projects - Third Edition

Jonathan Linowes

ISBN: 978-1-83921-733-3

- Understand the current state of virtual reality and VR consumer products
- Get started with Unity by building a simple diorama scene using Unity Editor and imported assets
- Configure your Unity VR projects to run on VR platforms such as Oculus, SteamVR, and Windows immersive MR
- Design and build a VR storytelling animation with a soundtrack and timelines
- Implement an audio fireball game using game physics and particle systems
- Use various software patterns to design Unity events and interactable components
- Discover best practices for lighting, rendering, and post-processing

Packt is searching for authors like you

If you're interested in becoming an author for Packt, please visit `authors.packtpub.com` and apply today. We have worked with thousands of developers and tech professionals, just like you, to help them share their insight with the global tech community. You can make a general application, apply for a specific hot topic that we are recruiting an author for, or submit your own idea.

Share Your Thoughts

Hi!

I am Jon, author of Augmented Reality with Unity AR Foundation, and Unity 2020 Virtual Reality Projects books. I really hope you enjoyed reading my book and found it useful for getting started build AR and/or VR applications. I think you're now ready to go out and augment your world!

It would really help me (and other potential readers!) if you could leave a review on Amazon sharing your thoughts on Augmented Reality with Unity AR Foundation.

Your review will help me to understand what's worked well in this book, and what could be improved upon for future editions, so it really is appreciated.

Best Wishes,

Jon

Index

www.ingramcontent.com/pod-product-compliance
Lightning Source LLC
Chambersburg PA
CBHW060923060326
40690CB00041B/3006